WISDOM *for* HEN KEEPERS

WISDOM *for* HEN KEEPERS

500 TIPS FOR KEEPING CHICKENS

CHRIS GRAHAM

The Taunton Press

The Taunton Press, Inc., 63 South Main Street
P.O. Box 5506, Newtown, CT 06470-5506
e-mail: tp@taunton.com

Conceived, designed, and produced by
Quid Publishing
Level 4 Sheridan House
114 Western Road
Hove BN3 1DD
www.quidpublishing.com

Library of Congress Cataloging-in-Publication Data
Graham, Chris, 1961-
Wisdom for hen keepers : 500 tips for keeping chickens / Chris Graham.
pages cm
Includes bibliographical references and index.
ISBN 978-1-62113-762-7 (hardback)
1. Chickens. 2. Hens. 3. Urban agriculture. I. Title. II. Title: Wisdom for henkeepers.
SF487.G6368 2013
636.5--dc23

2013029348

Printed in China
10 9 8 7 6 5 4 3 2 1

This is for Rachel;
she is my wisdom.

CONTENTS

INTRODUCTION

Much has been written about keeping chickens over the years, ranging from dusty old textbooks to online blogs and Twitter chatter. Inevitably, the internet is playing an ever-greater role as a source of information, with several established forums now providing decent help and support for their hen-keeping users. However, there's also a tremendous amount of incorrect information out there, peddled by "overnight experts" who create convincing arguments for all sorts of poultry-related no-nos. In contrast, the idea of this unique book is to provide an authoritative reference source for the novice hen keeper. The 500 practical, down-to-earth tips will guide readers through the basics of caring for their chickens properly, with consideration and with the greatest of pleasure.

GETTING STARTED WITH HENS

The popularity of keeping chickens in a backyard, be it in an urban or rural setting, has seen a real surge over the past decade. With an ever-growing awareness about eating and living healthily, more and more people are realizing that, quite apart from being a fun and interesting hobby, keeping a small flock of hens can make a worthwhile difference to what the family eats. However, hens are a commitment. Understanding the extent of the responsibility you're taking on can make all the difference to the ultimate success of the project.

BEFORE YOU BEGIN

TIP 1: *Think of the benefits*

As a hobby, keeping chickens is hard to beat. It's a thoroughly engaging pastime that's open to almost anyone with a yard and time on their hands. Initial setup needn't cost a fortune, and nothing holds a candle to the taste of a freshly laid egg. It's also a hobby that you can enjoy on many levels; at its most basic you can keep just three hens in a small backyard run for a supply of healthy, nutritious eggs. More involved options include hatching and rearing chicks, producing stock to a specific breed standard, or even exhibiting your birds in a poultry show. But, first things first: do think long and hard about what level you can realistically take on.

TIP 2: *Be prepared for hard work*

It's very important to appreciate everything that's involved in keeping hens properly. Too many people rush into the hobby, caught up on a wave of enthusiasm and a desire to become involved as quickly as possible. The temptations are certainly great, with many books and several magazines telling would-be keepers how easy it is to own hens, and what fun they are to have scratching around in the backyard. All this, of course, is true—up to a point. You shouldn't be under any illusion: there will be hard work involved. You'll need to invest both time and money in keeping your hens properly. Things will go wrong, and fixing them can be expensive and sometimes upsetting. So it's vital to appreciate the responsibility you're taking on, and to be aware that the novelty can wear off!

TIP 3: *Make sure you've got the necessary cash*

It may seem like an obvious thing to say, but plenty of people underestimate the cost of getting started with chickens. Buying the birds and all the paraphernalia needed to provide them with a safe and pleasant living environment is likely to cost several hundred dollars. This represents a significant investment for most people, and it should never be regarded as something you can do "on the cheap." What's more, you'll need the bulk of the money up front: it's necessary to get off on the right foot! If you are forced to buy things in stages then, for goodness sake, get all the hardware first, and the birds last. You'd be surprised by the number of people who rush out and buy their birds and then call in at the local poultry equipment supplier on the way home to pick up a chicken coop and some fencing. This, of course, is completely the wrong way to do things.

TIP 4: *Meet some chicken keepers*

Given that there's so much choice for the novice keeper with regard to bird breeds and all the hardware involved in their upkeep, it really does pay to do your research before taking the plunge. Those who fail in this respect all too often live to regret it. I know it's tiresome and requires patience, but getting a feel for the hobby before you start really can pay dividends. There's plenty of useful information on the internet and in books and magazines, but probably the best option is to get out there and meet existing keepers. You can do this by visiting an agricultural expo where there's a poultry tent, or finding the details of your local poultry club or 4-H group. These organizations are packed with enthusiastic keepers who'll be only too willing to offer help and guidance to those getting started.

ASSESSING YOUR LEVEL OF COMMITMENT

TIP 5: *Be realistic about your expectations from the start*

It's all too easy to rush into hen keeping on a wave of enthusiasm when the sun is shining and the birds are singing. However, things can feel a bit different once the weather turns and warm, cloudless days are replaced by wet and windy conditions. But it's vital to understand that chickens will continue to need daily attention, whatever the weather. You can't simply ignore them because there's a heavy frost, or it's snowing. Keeping poultry successfully is all about commitment, routine, and regular observation. If you're not prepared to invest the time and effort needed to undertake the essential husbandry work involved, then rethink your plans before it's too late.

TIP 6: *Make life easier for yourself by taking things slowly*

Start slowly with your new hobby, however tempting it may be to buy that super-sized chicken coop and those 15 pretty birds. It makes a lot more sense for beginners to keep everything at a very manageable level for the first season. You'll be far less likely to run into problems of any sort with a small chicken coop and just three hybrid layers. It's much better to make sure you're going to enjoy keeping chickens at home by not over-committing yourself from the outset. It's far less daunting, and you'll find every aspect of the process easier, quicker, and more affordable by starting small.

TIP 7: *Plan sensibly for your family's needs*

It's all too easy to imagine that it'll be much more fun to keep six birds rather than just four. This can be the case, of course, if you have the space and facilities to accommodate them. However, one aspect that many beginners fail to appreciate is the number of eggs likely to be laid. If you're considering modern hybrid hens—or a decent, utility strain of pure breed—don't forget that these birds in their prime will lay five or six eggs each, for every week during the season. Are you able to cope with 36 eggs a week, every week for the majority of the year? Even four hens are likely to produce 24 eggs a week, which is normally quite sufficient for most domestic kitchens.

TIP 8: *Consider your lifestyle*

The time to be indecisive and have doubts is before you get the birds. It's important to be absolutely sure that you want to keep chickens before you actually take the plunge. Be realistic when assessing your own lifestyle, and how taking on a few hens will change things. If you live the kind of carefree existence that involves jetting off for last-minute road trips or simply because you saw a deal on the internet, then you need to think seriously about how keeping poultry will cramp your style. Chickens need attention, and they need it every day, 7 days a week, fifty-two weeks a year. Imagining that you will be able to get away with less input than this and believing that, somehow, things will be all right just isn't fair for the birds you're thinking about buying.

TIP 9: Act in haste, repent at leisure

The popularity of keeping hens in the backyard nowadays means that nearly everyone knows somebody who is already involved with the hobby, so there's no shortage of anecdotal advice about getting underway with your venture. By its very nature, though, most of this tends to be extremely positive and encouraging. You'll hear few horror stories about cases where things went wrong, which means that it can be all too easy to fuel your own sense of enthusiasm for starting before you're truly prepared. Do bear in mind, though, that keeping chickens isn't a suitable hobby for all. It involves regular work, considerable commitment, and, a routine. If you only discover this once you've got your birds, then looking after them can quickly become a real chore.

TIP 10: Friends can help, but...

While it's always handy to have friends and relatives who can step in when there's an emergency, or you aren't able to be with your birds for some reason, never take this sort of assistance for granted. Occasional helpers won't know your birds like you do and, even if they are experienced poultry keepers themselves, they may well miss important, tell-tale signs of trouble that you would have recognized. Of course, it may be that the neighbors you entrust with feeding and watering your birds while you enjoy a three-day weekend at the cabin haven't got the first clue about caring for chickens. Will they have listened properly to the instructions you gave and can they be trusted to do what's required when it needs to be done? You need to think very carefully before passing the responsibility for your birds on to others.

GETTING A GOOD START

TIP 11: Start in spring

It's far easier to get your bearings with a new group of chickens when the weather's fine than when it's cold, windy, and wet. The smart thing to do is to buy your birds early in the spring to maximize the amount of good weather in which you have to become accustomed to caring for your feathered friends. If you're getting point-of-lay pullets (young hens that are on the verge of egg production), it's also worth remembering that birds at this age will grow best with the sun on their backs. Strong, healthy development is enhanced by good weather, and birds who mature time of year—when insect life is at its height—will be able to get the best possible start to their adult life, having done all the important growing before the weather turns.

TIP 12: Space is a key factor

While just a few hens won't require much run space, it's very important, from a welfare point of view, that they do have enough. One of the worst things you can do to chickens is overcrowd them, so it's essential that the number of birds you keep is appropriate to the space you have available for them. While surprisingly resilient in many ways, hens are susceptible to stress. This can be caused by all manner of triggers, and overcrowding is certainly one of them.

TIP 13: Space isn't everything, though

At the risk of contradicting the previous tip, don't worry if you don't have a huge backyard. While you'd ideally give a group of hens as much outdoor space as possible—so that they can enjoy a "free-range" lifestyle and supplement their regular pellet- or mash-based diet with tasty insect treats from the garden—there are cases where this simply isn't possible, and it is perfectly fine to keep a few laying hens in a confined space if you look after them properly. We'll explore some options for keeping chickens in smaller spaces in chapter 5 (see tip 204). However, the most important thing to bear in mind at this stage is that you should avoid forcing confined birds to live on a muddy, droppings-contaminated surface.

TIP 14: Tend little and often

Chickens require regular attention to keep them fit and well. They benefit in many ways from frequent interaction with their keepers, something that's important for establishing mutual trust and ensuring a decent level of manageability. If you're contemplating keeping hens for the first time, you should be aware that the birds will need attention at least twice a day: to check their overall condition as well as make sure they have both food and water available. Straightforward observation is as much a part of good poultry keeping as anything else. By keeping a regular eye on your birds, you'll quickly learn about their individual characters, behavior, and how they interact with each other. Also, and just as importantly, you'll develop the ability to spot trouble, and nip potential problems in the bud.

TIP 15: Establish a routine

One of the fundamental secrets to success with chickens is establishing a daily routine. It may sound boring in these days of flexible lifestyles, but keeping chickens in your yard is all about routine. Hens are creatures of habit; they like the humdrum regularity of a fixed husbandry pattern. So letting them out and shutting them up at predictable times, and feeding the same food at the same place each day, represent little short of chicken heaven for the birds. Their habitual nature means that your birds will get used to those who regularly look after them. So, for the most consistent results, maintain as much continuity as you can. While impressively resilient in many ways, chickens can be extremely vulnerable to stress, and this is something that can be triggered by apparently trivial events such as feeding them a strange food at the wrong time.

TIP 16: Stretch beyond the nine-day-wonder effect

As with anything new, there's always the risk that the excitement of having a few chickens in the yard might wear off sooner than you might have hoped. Compared with an Xbox or PS3, a few hens scratching around in the yard can seem a bit dull to young eyes. However, it's important that you persevere, doing all you can to keep interest levels up. So, whether it's setting up a webcam inside the chicken coop, creating charts for tracking eggs laid, or simply encouraging the hens to take feed pellets from the hand, children can learn a tremendous amount from an active involvement with chickens. Chickens can provide a valuable introduction to the cycle of life and death, lessons about animal behavior, and also an essential insight into the link between a good environment and a healthy, productive life.

TIP 17: Involve the family to keep everyone interested

Plenty of newcomers to the poultry keeping hobby are encouraged to get involved by their children, who typically find being around chickens and collecting eggs absolutely fascinating. However, it's important to do all you can to maintain interest levels throughout the family, unless you want to end up doing all the work yourself! If everyone can be assigned jobs and given a little responsibility, then so much the better. Children, especially, love to feel included. Encouraging them to handle the birds properly, on a regular basis, is a great thing to do. It'll build confidence in both them and the birds, and you'll be able to start teaching the youngsters about how to assess condition and general health when handling birds, skills that'll serve them well when they graduate to having their own flock!

TIP 18: Join your local poultry club or 4-H group

You can read all the books, magazines, and internet articles you like, but, with a specialist hobby like poultry keeping, there's no substitute for hands-on experience. Of course, beginners inevitably have to start at the bottom of a steep learning curve, and first-hand knowledge and experience must be gained at every opportunity. This is where your local poultry club or 4-H group has an incredibly valuable role to play. Unless you live in the middle of nowhere or in a large urban area, there's bound to be a local chapter, and it'll be packed with enthusiastic souls just dying to pass on their poultry keeping wisdom. These organizations represent skills and resources that every novice should plunder. They're cheap to join, and the benefit potential offered by the many kindred spirits you'll find is enormous.

TIP 19: *If you're feeling overwhelmed, ask for help*

It's a common mistake among novice keepers to underestimate the amount of work required to care properly for their birds. Sometimes, even with the best will in the world, things can get a little out of hand; the birds might start fighting with each other; they could wreck the area they're in much quicker than expected; or it may be that the hens simply start going downhill for no apparent reason. Whatever the problem, though, it's important that you do something positive about it. Never forget that the birds in your care are completely dependent on you for their health and well-being. So if things start going wrong—or just don't seem right—then don't hesitate to ask for help. Consult your vet (preferably one with poultry experience) or seek the advice of a knowledgeable and experienced friend or fellow club member. The worst thing you can do is nothing at all.

TIP 20: *Ultimately, never be afraid to throw in the towel*

You may decide that chickens simply aren't for you. There's no shame in this, but the important thing is to act on that decision sooner rather than later. The nightmare scenario, as far as the birds are concerned, is that they start being neglected. This may be completely unintentional as far as you're concerned, but a loss of interest—or the realization that you simply don't want to be a chicken keeper after all—usually manifests itself in declining attention being paid to the birds. The worst thing you can do is allow them to suffer due to erratic feeding, an increasingly dirty environment, or a lack of fresh, clean drinking water. So don't delay if you want to give up. Be responsible about finding the birds a safe and caring new home, and recoup some of your financial outlay by selling your chicken coop and other equipment on the internet, or via your local poultry club.

KNOWING THE BEST ENVIRONMENT

TIP 21: Be realistic about capacity

One of the golden rules of poultry keeping is that you should never overcrowd your birds. Packing too many into a space that's too small for them will trigger all sorts of undesirable behavior, the most serious of which is probably cannibalism. This, as you can imagine, is distressing for the victims as well as the keepers, especially if there are children in the household. Be under no illusion that chickens can turn very quickly from mild-mannered, grass-scratching cuties into fiercely aggressive killers when provoked. Overcrowding can be one of the main triggers for this unpleasant behavior; it takes only a frustrated peck from one bird to cause a bleeding wound in another, and the rest of the flock will turn on the unfortunate, injured individual. Nine times out of ten—assuming the keeper isn't able to intervene—there's only one outcome, and it isn't a good one.

TIP 22: Allow enough room to change locations

If you favor a chicken coop with an attached run that'll help keep your birds both safe and contained, it's important that you have sufficient space to move this unit around on a regular basis. Ideally, these coop/run combinations should be sited on shortish, fresh grass and will need to be moved every other day to prevent any permanent damage being done to the grass roots. The fact that the run needs moving so frequently should give you a good indication about the speed at which the birds can affect the surface and, obviously, the more birds there are for a given area, the quicker they'll have an effect. But it's not only saving the grass that's of concern; allowing the birds to turn the surface into a muddy mess will greatly increase the chances of exposure to disease and parasites.

TIP 23: Not all chickens are garden-friendly

While watching a group of hens scratching contentedly on the lawn in the warm summer sun is a universally appealing image, it's important to appreciate that the birds aren't simply scratching around for the fun of it; they're engaged in an almost-constant search for food. Worms, grubs, insects, and most forms of seed—as well as grass and other vegetation—will all be on the menu. The birds will dig as far as they need to find these treats, and certainly will pay no attention to a carefully crafted garden layout. Consequently, it can take only a handful of hens to strip, rake, and ruin a well-cultivated herbaceous border. However, the tendency to do this does vary from breed to breed, so it is possible to minimize the destruction.

Keeping feather-legged breeds, such as the Silkie, the Brahma, or the Cochin (see chapter 3 for more on chicken breeds), is a good idea for avid gardeners, as these will tend to scratch less than clean-legged birds. Bantam versions of the pure breeds can be a sensible choice, and the diminutive true bantams, such as the Belgian, Serama, and Dutch Bantam, are good, too.

TIP 24: Savvy vegetable growers will have great compost

It's not all bad news for gardeners. Those who like growing vegetables can really benefit from the activities of chickens. A group of birds put on to a harvested garden patch will do a fabulous job of clearing the ground of slugs, snails, and any leaf and root matter that may remain. What's more, as a byproduct of this ecofriendly operation, the droppings they produce as they work will do much to replenish the soil. Also, don't forget that the used bedding that you'll be removing from the chicken coop on a regular basis will make an excellent addition to the compost pile, helping to ensure the production of a rich and nourishing soil conditioner that'll work wonders on both vegetable and flower borders around the garden.

TIP 25: *Damp and soggy areas are best avoided*

It's common sense really, but chickens and wet, muddy conditions aren't great bedfellows. Poultry runs should never be located in areas where the ground has a tendency to flood or does not have good drainage. Bantams with short legs, and those breeds with feathered legs don't cope well with wet conditions; their feathers become sodden and mud caked, which can make it difficult for them to walk. Even clean-legged breeds can be adversely affected by the build up of mud on their feet, making it painful to stand and roost. Then there's the problem of all the mud and moisture that inevitably gets carried inside the chicken coop. This can cause a whole new set of problems if not dealt with, as a damp environment inside the chicken coop will seriously increase the chances of respiratory problems among the birds.

TIP 26: *Chickens feel safest with overhead cover*

It's thought that all domestic chicken breeds descend from the Red Jungle Fowl, which is a native of southeast Asia (eastern India, southern China, and Indonesia). The fact that this is a jungle-dwelling bird is an important one, as it has instilled its descendants with the inclination to feel happiest and most secure from sky-borne predators when under overhead cover. So chickens that are able to free range in a woodland environment are likely to be the happiest of all. Now, while most of us don't have the luxury of our own woods for hens to roam in, fruit trees, hedges, and small shrubs can all be used to good effect in the urban yard situation. But if you can't site your chicken enclosure around any of these natural forms of cover, then you should think about building a simple shelter or raising the height of the chicken coop so that the birds can take cover under that. (See also tip 179.)

TIP 27: *Snuffles and sneezes spread diseases*

Watching chickens go about their daily business is therapeutic in itself, but it also offers an important insight into their general behavior and overall health. Regular handling of the birds is better still, as it gives the keeper the opportunity to spot problems and nip them in the bud. For example, it's much easier to identify a bird suffering with watery eyes or nostrils—or that's making wheezing sounds as it breathes—while it's being held. These are important signs of potential respiratory problems that may require veterinary attention and treatment (see chapter 6). Conditions like this can also be extremely contagious, so it always makes sense to isolate birds found suffering in this sort of way.

TIP 28: *Hens are generally resilient, but there are limits*

In contrast to their often delicate appearance, chickens are actually amazingly tough creatures, and can cope well with all sorts of upsets. However, when trouble does strike, these birds are masters at disguising their health problems during the early stages, and this can make it difficult for novice keepers to spot when things are going wrong. There will be signs, of course, but they'll be subtle. The situation can be compounded by the fact that once things eventually become obvious—the suffering hen becomes lethargic, stops laying, appears hunched and fluffed up—the chances are that whatever's causing the problem is already well advanced and veterinary attention will be required.

TIP 29: *Keep the coop and run clean for the neighbors' sake*

Never forget that although you and the rest of your family will be full of excitement at the prospect of getting your first chickens, the idea might not be quite so appealing to your immediate neighbors. Those next door may be all too quick to focus on the potential negatives, with rats, smells, and noise being the three most common objections thrown up by the doom-mongers. While all three "undesirables" can occur, chickens that are kept properly shouldn't suffer with any of them. Rats will only be attracted if there's plenty of spilled or badly stored feed and runs that are not properly secured. Smells won't be created if the coop and run are kept clean, and noise shouldn't be an issue if you don't keep roosters.

TIP 30: *It might take only one complaint to the authorities...*

Unfortunately, the way things are nowadays, it's common for local authorities to take the side of the complainant in the case of domestic chickens. The presence of rats is frequently blamed on chickens when, in many cases, the pests are living in compost piles or under log piles, and feeding on badly packaged food waste or the treats others put out for wild birds or pet waste left in the yard.

UNDERSTANDING YOUR HENS

TIP 31: A chicken's needs are essentially simple

One of the great things about keeping poultry is that it's a relatively straightforward thing to do. There aren't lots of complicated procedures to learn, or things that need to be practiced over and over again, as with many other hobbies. Instead, it's as much about common sense as it is methodical attention to detail. In reality, chickens kept in a backyard environment don't have complicated needs; all they require is secure, dry, and draft-free night-time accommodation, effective protection from predators, an adequately sized run area, plus a regular supply of clean, fresh drinking water and food. Meet these basic needs and your hens should be happy as well as productive.

TIP 32: Provide a sense of freedom

Space is always a key factor with chickens, whether you have three birds or 300. It's essential that the hen coop is large enough for comfortable roosting at night and that, during the day, the birds have plenty of space to stretch their wings and participate in natural behavior. While most decent coop manufacturers will specify the maximum capacity of the units they sell, you can double check this by measuring the total roost length available inside. As a guide, allow 8in (20cm) of roost space per large fowl bird (a little less for bantams). As far as run size is concerned, it can never be too big! If space is limited, and you're keeping the birds in a fixed unit, then pay plenty of attention to floor litter condition and provide lots of diversions to prevent the occupants from getting bored and unruly—something you do not want among hens confined in a small space.

TIP 33: Hens are surprisingly vulnerable to stress

A very important part of the art of good poultry management is to ensure that the stress levels among the birds are kept to an absolute minimum. Now, this may sound rather absurd, especially given that stress is such a "human" problem nowadays. But, rest assured, chickens can suffer with it, too, and, when they have it, not only will it affect their laying performance, but it can also lower their resistance levels to disease, making them far more vulnerable to ill health. Even the effects of internal parasites, such as worms—which may be completely manageable for a healthy bird—can become a seriously debilitating problem for a stressed hen. The trouble is, stress among chickens can be triggered by many factors; overenthusiastic children, inappropriate handling, a persistently barking dog next door, erratic feeding patterns, and even fear of predator attack.

TIP 34: You can tell an awful lot from basic behavior

Chickens should naturally be active and inquisitive characters. Healthy stock will be permanently on the move, scratching and pecking at whatever piques their curiosity. Consequently, it's a pretty safe bet to assume that birds which appear lethargic and uninterested will be suffering from something; all is not well! So if you are visiting a supplier to buy birds, avoid those that are standing quietly away from the rest of the flock, with their heads pulled in close to their bodies and their feathers slightly fluffed.

TIP 35: Know the pecking order

It's no myth; pecking orders really do exist within groups of chickens, and it's serious business. The bird at the top certainly rules the roost, and will have exerted its dominance on the others by force, if necessary. Those at the lower end of the scale most definitely know their place, and tend only to move up the order if new birds are added to the group. The existence of the pecking order is why it can be tricky introducing "strange" birds to an established group. A new order will need to be established before things can settle down again, and this can, on occasions, involve some unpleasantly aggressive behavior. Keepers always need to be vigilant for the first few days after an introduction and should be prepared to offer isolation to any bird that starts to suffer or gets injured.

TIP 36: Observe and act on cruelty within your flock

Even the most docile of hens can turn into rabid killing machines! Witnessing how your normally friendly backyard layers deal with worms, frogs, or even a dead mouse if they happen to find one, should provide a clue about their potential for carnivorous behavior. Problems arise if one of their number becomes injured. A bleeding wound is like a red flag to a bull as far as its flock mates are concerned, and they'll start pecking at the problem mercilessly; the color of the blood appears to be the trigger. In the worst cases, this will continue until the victim is pecked to death, which can be completed in a matter of hours. The same can happen following the annual molt, when the new feathers are just emerging and have a visible blood supply at their base. This is why regular observation of your flock is so important; timely intervention can literally save lives.

TIP 37: *You can't expect an egg every day, forever*

A fit and healthy young hen won't lay seven eggs a week; she simply isn't capable of that. Most good hybrids will manage five or six over a seven-day period, but pure breeds will generally lay fewer than this. Breeds that have been selectively bred over the generations for exhibition purposes—with particular emphasis on profusion of feathering—won't lay many eggs at all. Keepers of some Orpingtons, for example, are lucky if they see 50 eggs a year from their hens. The Mediterranean pure breeds, however, such as the Leghorn, and the recognized utility breeds including the Rhode Island Red and Sussex, still produce worthwhile numbers of eggs, although even these are not in the hybrid league.

TIP 38: *Good health and welfare are essential for egg production*

You can't simply expect your hens to lay eggs; factors such as age, health, and environment all play key roles in the birds' ability to produce a regular supply and perform to their full potential. Stressed birds won't lay well and neither will those that are being overfed and are fat. Hens kept in a poor environment, that feel unsafe or are being bullied will lay erratically or not at all. Diet is a very important factor too. It needs to be properly balanced, containing the right mix of protein, vitamins, and minerals to ensure that both the egg's contents and its shell are properly formed. Shortfalls in nutritional balance can lead to shell quality issues (sometimes it's soft or even nonexistent), which is why most keepers opt to feed their birds with commercially produced layer pellets or mash.

TIP 39: Eggs or feathers, but not both

Plenty of new keepers are taken by surprise when their birds first start the molt. Not only is the rapid loss of feathers quite a shock to behold (many novices automatically assume their birds have caught some terrible disease), but this annual event coincides with a halt in egg production. too. With both eggs and feathers having a very high protein content, chickens can only produce one or the other at a time. Understandably, at the end of the breeding season and the approach of autumn, it's feather replacement that takes priority, ensuring that a fresh new set is in place before the temperatures start to fall. The break in egg production at this time also gives the bird's body a welcome rest.

TIP 40: Egg-laying has a finite limit for every chicken

Interestingly, each chicken hatches with the potential to lay a predetermined maximum number of eggs, although not many live long enough to ever reach the end of production. Modern hybrid hens have been bred for maximum production in the first eighteen months or so of their lives, after which the number of eggs laid starts to fall away dramatically. Consequently, commercial laying flocks have a working life of no longer than this. The performance of utility pure breeds is rather more measured, without the dramatic, initial rush and with a slower reduction in numbers as time passes. So, keepers of good strains of breeds such as the Wyandotte or New Hampshire Red can expect a slower but steadier egg production rate over three or four years.

THE BENEFITS OF KEEPING HENS

TIP 41: Enjoy the outdoors experience

The idea that you can temporarily lose yourself as you care for your birds is an important one. The time it takes you to fill the feeders and waterers each day, or the hour or so you set aside at the weekend to clean out the coop, can represent great contemplation time for those with a busy schedule; this can be an important oasis of calm within the organized chaos of our increasingly hectic lives. Even the few minutes spent outside as you carry out your husbandry chores is beneficial. Fresh air and a bit of exercise is never a bad thing, especially now that so many of us spend every waking hour staring at a computer screen.

TIP 42: Indulge in egg-cellent food

Most people get into the poultry hobby because they want to benefit from better eggs than they can typically buy at the supermarket. While there are certainly better choices available on the shelves these days, eggs produced at home by healthy hens that enjoy the space to range freely are still superior to most. And, of course, a good egg is one of the most complete foods you can eat; high in top-quality protein, rich in vitamins and minerals, plus a great source of mono- and polyunsaturated fats. The protein found in an egg is so good that it sets the benchmark against which all other foods are assessed.

TIP 43: *Unwind with your new feathered friends*

In the same way that watching tropical fish swimming around in a well-stocked tank can be therapeutic, so can enjoying the antics of your backyard hens as you sip from a mug of coffee or glass of wine. While not scientifically proven as far as I'm aware, there's plenty of anecdotal evidence suggesting that simply taking some time out to watch the perpetual motion of your chickens nourishes both the mind and soul. So time spent with your birds is never wasted. After a hard day at the office, or a stressful time running your children here, there, and everywhere, twenty minutes losing yourself in their world can be just what the doctor ordered!

TIP 44: *Regenerate your garden patch for free*

If, like many people with limited garden space available, you grow your vegetables in raised beds, then it's a smart idea to make a wooden and chickenwire framework that'll fit over these beds. This will allow you to contain your hens over one bed at a time so that, once cropping has finished, they can pick away at the vegetation left and snack on the grubs, worms, and slugs they are bound to find there too. Clearing the ground of pests in this ecofriendly manner is a great thing to do; the soil will be conditioned and fertilized at the same time as well. Make sure, though, that you haven't used anything toxic on the beds before allowing the chickens access. Slug pellets, other pesticides, and some chemical-based fertilizers could be harmful to the hens, so take care.

TIP 45: *You can rear for roasting*

The chicken is a strange bird, in that it bridges the usual divide that exists between domestic animals and livestock; it's a productive pet. But if there's one thing that's guaranteed to divide the hobby it's the great "to eat or not to eat" debate. While there are a great many keepers who wouldn't dream of killing their birds for the table, an increasing number are starting to appreciate the benefits of growing their own meat birds. As concerns about animal welfare continue to rise, and consumers become ever-more aware of what they and their families are eating, the idea of preparing birds for the table that have been reared happily, naturally, and slowly at home is gaining popularity.

TIP 46: *Don't forget the conservation angle*

Many of the pure breeds are struggling nowadays in terms of overall numbers. Dedicated groups of enthusiasts work hard to preserve whatever stock is left but, in some cases, this is a real struggle. Some of the more fringe breeds—which don't have their own, independent clubs for support and promotion—are literally teetering on the brink of oblivion, and continue to be in desperate need of more help. More keepers are needed to help prevent some of these strains from being lost completely, and this is really important work that can take your poultry keeping to a whole new level. The beauty of it is, though, that it's not expensive to get involved with. Despite the rarity of the most endangered breeds, stock and hatching usually remain very affordable; the biggest requirement when getting involved is patience.

TIP 47: *Breed to the standard; that's the key*

If you start keeping male birds and become interested in breeding then it's important, when working with pure-bred stock, to familiarize yourself with the relevant breed standard. The appearance of every established pure breed is dictated by the description given in the breed standard: a set of agreed-upon and approved guidelines that summarize the way a bird should look, in terms of feather type and color, body size and shape, leg, beak and eye color, comb type, and tail size, etc.

In the United States, the American Poultry Association (for website details see p. 288) oversees the breed standards, and publishes the *American Standard of Production* book, essential reading for anyone serious about breeding properly. Many of the breed standards were established back in the mists of time, and the countless generations of enthusiasts that have followed them carefully are the reason why we can all still enjoy such a wonderful variety of distinct chicken breeds today.

TIP 48: Socialize with other hen keepers

Yet another significant benefit of becoming actively involved in the poultry-keeping scene is that there can be a great social side to enjoy. Not only will fellow poultry club or society members be able to offer valuable advice on all aspects of the hobby, but they can become great friends too. The nature of this hobby tends to mean that it attracts kind and compassionate characters with an interest in rural affairs and animal welfare. On the whole, they're a sociable lot too and, with so many shared interests, there's usually plenty to talk about. There's great camaraderie within poultry exhibition circles as well. Obviously, the competition at shows is usually pretty fierce but, once the judges have made their decisions, everyone relaxes to enjoy the social side of the event.

TIP 49: *Turn your hobby into a part-time business*

Given that poultry keeping is so popular nowadays, increasing numbers of people are looking at the pastime and seeing a business opportunity. Already it seems as though the internet is awash with specialist breeders and coop and equipment suppliers but, it has to be said, the quality of the birds and equipment being sold does vary enormously. So, like any other type of business, success with a poultry one will hinge on the quality of product and the efficiency of service provided. People's expectations are rising ever higher, so you'll need to know your subject and deliver the goods, otherwise your reputation will be destroyed before you know it. Unlike twenty years ago, negative feedback can now be posted in the blink of an eye, for all to see.

TIP 50: *Keep it environmentally friendly*

Keeping chickens is a good thing to do on many levels but, especially nowadays, we'd do well to remember just how environmentally friendly it is, too. Quite apart from being fun, interesting, and educational for children, the supply of eggs (and possibly meat) from the birds can contribute to cutting your household's food miles. Also, given sufficient space in which to free-range, chickens will usefully supplement their regular pellet or mash diet, helping to minimize the amount of commercial feed they consume. What's more, the litter coop and nest box bedding material will all be biodegradable, making it an ideal addition to the compost pile. Even the cost of electrical additions such as light bulbs, automatic door closers, or heated waterers can be kept as low as possible by the use of a battery-powered system aided by a solar panel.

HEN-KEEPING EQUIPMENT

Successfully looking after chickens requires a mixture of basic knowledge, common sense, and the right equipment. It's not rocket science, but using good tools makes life easier. Unfortunately, with poultry keeping being so popular, the market has been swamped by equipment suppliers. While this is great as far as choice and price are concerned, it has led to an increase in outlets peddling substandard and poorly made products. So it's important to pick your supplier carefully, to buy on recommendation, and to avoid skimping on what you pay.

FOOD AND WATER

TIP 51: *Be mindful of where and what you scatter*

The appealing vision of scattering corn across the farmyard gives rather a false impression about what is needed to feed modern hybrid hens. Hens need top-quality poultry ration to maximize their laying performance; either layer pellets, crumble, or mash. Feed of this sort will contain everything a bird needs to be well nourished and productive in terms of eggs. Consequently, a good layer ration should be sufficient, although many keepers also persist with feeding "treats," which, in some cases, do more harm than good.

TIP 52: *Premium poultry feed brands offer top quality*

It may seem obvious but, as with so much else in life, you get what you pay for when it comes to poultry feed. So the product that's produced by an established feed mill working to an agreed upon set of industry standards is likely to provide a better quality of feed than a smaller operator that may be cutting corners to keep prices as low as possible. In terms of the brand to choose, if you can't decide yourself, then take the advice of a keeper. Buying on recommendation like this is often the best way to go.

TIP 53: *Avoid buying too much feed at once*

When ordering feed, think about the rate at which your hens are consuming it, so that you don't order too much at any one time. The last thing you want is for it to start going moldy before you're able to use it. It can be tempting to opt for the savings associated with a larger order, but this will count for nothing if you end up having to scrap a proportion of what you buy because it has spoiled.

TIP 54: *Using a well-designed feeder will save you money*

The price of poultry feed is on the increase and, apart from the hen coop, it represents one of the biggest costs involved in keeping chickens at home. Feeder design can play a crucial role in preventing unnecessary wastage. The primary objective should always be to ensure that the expensive poultry feed you provide is eaten exclusively by your chickens and not by local wild birds, mice, squirrels, or rats that may be passing. Using a properly designed poultry feeder (with rain cover and legs or elevated on a cinder block) will keep the feed in better condition than if it is in an open bowl on the ground and will also help stop the feed from being pilfered by uninvited diners.

TIP 55: *Take care with the storage of feed*

A common mistake that plenty of beginners make is to fall short of the mark in terms of feed management. Layer pellets have a "best before" date, which it's important to adhere to. The danger is that the fat and oil in old feed start to deteriorate, which not only destroys the vitamin content but also starts to turn the whole lot rancid; the chickens simply won't eat it. Poultry feed will need to be stored in a cool, dry environment, where it's safe from rodent attack. The paper sacks it is normally supplied in are no deterrent to squirrels, mice, and rats, so don't leave them standing on the ground under a lean-to or in the garden shed. A sensible measure is to put the feed into a metal trash can with a tight-fitting lid, where it's much safer. If possible, store this out of direct sunlight to avoid wild temperature fluctuations.

TIP 56: *Make sure all the birds get their share*

The poultry pecking order always has an influence at feeding time. The most dominant birds will always have their fill first, while those at the opposite end of the seniority scale will usually have to wait their turn. In some instances, though, if a bird is being badly bullied by others in the group, then it may be prevented from eating. This, obviously, is a problem that you—as their keeper—need to watch for and act on before the situation gets too serious. One simple idea that usually solves the problem is to introduce another feed station (or maybe two more), so that the bullied bird has a chance to get away from the others and eat on its own.

TIP 57: *Drinking water is essential*

Never forget the vital importance of maintaining a constant supply of fresh, clean drinking water for your chickens. Dehydration is a very serious problem in chickens, and is something that can strike surprisingly quickly if drinking water needs are neglected. A bird can lose all of its fat and recover, or half of its protein and still survive. However, if it loses just one-tenth of its body moisture, then it will die. So it makes sense to use waterers with a large reservoir of water, especially if you have to leave your birds unattended for hours at a time during the day.

TIP 58: *Remember, cheap waterers and feeders won't last*

In most cases, cold weather spells the end of the working lives of cheap feeders; their thin plastic hardens and becomes brittle due to its exposure to light and sunshine, and then, when the winter frosts arrive, cracking is almost inevitable. Legs can shear off plastic feeders with even the most minor provocation, and water that's allowed to freeze and expand inside a plastic waterer will split the sides with ease.

TIP 59: *Clean waterers and feeders prudently and regularly*

The vital nature of good feed and water supplies necessitates that you pay particular attention to the condition of the containers delivering these essentials. Both waterers and feeders can become contaminated quickly, although some designs are a good deal worse in this respect than others. Feeders such as the traditional trough, which are open to the elements and allow the hens to stand in them as they eat, are obviously prone to getting much dirtier than modern units that are raised off the ground and have protected feed dispensers. Waterers get slimy and dirty quickly too, simply through everyday use and so, like feeders, they require careful and regular cleaning using an acceptable, nontoxic disinfecting agent. Many keepers use mild soap and water to clean and a vinegar solution to disinfect, as these are both safe and effective when used properly.

TIP 60: *Apple cider vinegar prevents mold*

It's not only the lip or bowl of a poultry waterer that can cause problems; the inside of the reservoir tends to develop green algae if given the chance. Some of the bottle-type waterer designs can be difficult to clean because the filler necks aren't wide enough to get a hand inside, so you end up having to do your best using a long-handled brush. However, one way of helping to keep this kind of algae growth in check is to add a dash of apple cider vinegar (ACV) to the drinking water. This will lower its pH—making it slightly acidic—creating an algae-hostile environment that prevents its buildup in the first place. Incidently, ACV is a good supplement for the birds as well, acting as an effective gut cleanser. However, do not use this product in water destined for galvanized waterers, as the acidic solution reacts with the metal.

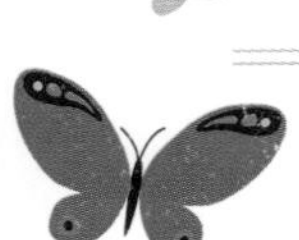

KEEPING YOUR BIRDS SAFE

TIP 61: *Keepers should take their duty of care seriously*

Part of the duty of care you have towards your chickens requires you to protect them from danger and, nowadays, that can come carried on two or four legs. Depending on where you live, foxes, raccoons, coyotes, opossums, weasels, mink, dogs, skunks, and rats can all pose a very real threat to your birds. But attack can come from the air too, with hens (and their young) in rural areas being at risk from birds of prey. Raccoons, coyotes, opossums, and crows can become habitual takers of eggs from the chicken coop, once they learn the trick. The other two-legged predator to bear in mind is your fellow man, so it really is a case of "keeper beware!"

TIP 62: *Chickens can get spooked by children and pets*

What chickens enjoy best is a routine, predictable lifestyle. They hate anything out of the ordinary, and can get stressed by sudden shocks or erratic behavior. Consequently, excitable children or pets (especially dogs) can easily spook hens. Just about the worst thing that can happen is for a young child to get in with the birds and to start rushing around, chasing them. If they are hens that you've been gradually training to become more friendly and easily handled, all this work can be undone in a matter of moments in the company of a rampant child, and take even longer than before to sort out. Regaining the confidence of birds that have been shocked into mistrusting humans is no easy task.

TIP 63: *Chickens and rats go hand in hand*

Part of the rat problem is caused by failures on the keeper's part. Careless feeding, bad food storage, and rat-friendly surroundings can all work to increase the likelihood of a problem. Food spillages, over-filling feeders so there's feed available constantly, and failing to deal with nearby piles of rubbish, log stacks, and undisturbed compost piles will all attract these clever scavengers. What's more, left unchecked, what starts as a few rats seeking shelter and a free meal can all too quickly develop into a thriving colony that'll probably need to be professionally removed.

TIP 64: *Watch for the tell-tale signs of rat activity*

Rats will always do their best not to be seen by humans. The most obvious signs are the gnawed holes in the floor and lower sides of wooden chicken coops. These may be hard to see until all the floor litter is removed during cleaning. Black, pellet-shaped droppings are another sign, as are small holes in the ground, especially if found under the hen coop, beneath an old stack of logs, or hidden behind rarely moved garden machinery. Take care when investigating this sort of problem, and wear gloves at all times.

TIP 65: *Don't make your run a welcome refuge for rodents*

It's important to keep the area in and around your poultry pen clean and tidy. Clear away rubbish, piles of logs, dead leaves, and anything else that might offer a potential shelter for pests. Even the ground under paving slabs used for paths needs to be checked. Rats and mice are smart operators, and it's surprising how easily they're able to squeeze into and under things in search of food or a home. What they don't like is bright light and frequent, noisy activity. Rats especially are inherently suspicious creatures—one of the very reasons why they remain a most effective survivor in nature.

TIP 66: *Rodents pose a threat to hens and humans*

Although the best advice is to mount all chicken coops on legs, raising the structure at least 8in (20cm) above the ground so there's plenty of light underneath, lots of keepers continue to sit their chicken coops directly on the ground. Rats love this; the shelter, the warmth, the security, and the prospect of food nearby all make for the perfect refuge to colonize. Once installed under the coop, the rats will gnaw their way inside, through the floor, to start causing trouble. They can chew off tail feathers as birds roost and have been known to kill and eat young birds. Then, of course, there's Weil's disease, which is spread to humans when rats' urine-contaminated water enters the body, most commonly via unhealed cuts or the eyes.

TIP 67: *Be proactive with your antirodent measures*

Always assume that there is some degree of rat activity, and lay poison. It's best to use a commercially produced bait station, which shields the poison from other creatures you wouldn't want eating it, such as wild birds and pets. Bait stations need to be sited carefully (usually at the base of fences or under hedges where you suspect rat activity) and work best if smeared with earth to disguise their newness.

TIP 68: *Always take great care when using poisons*

You should take the advice of your local supplier about which poison is the most suitable for your situation and degree of problem. It's also worth noting that most rat poisons available for domestic applications work with a cumulative effect, so it's important to keep replacing the dose and do this until the time comes when the poison stops being taken. In reality, this is temporary respite and antirodent measures should never be totally relaxed.

TIP 69: *Foxes are cunning, hungry opportunists*

Traditionally, it's the fox that most people regard as the number-one threat to their chickens. It's a common misconception that foxes, once inside a chicken run or shed, simply kill everything in sight just because they like it. What's actually happening is that the fox is simply engaged in stage one of the feeding process. This, quite naturally, involves killing as many birds as it conveniently can. Stage two is removal, when the carcasses are dragged or carried away for immediate consumption or burial elsewhere. But this part of the process takes time, and the devastation is usually discovered by the keeper before the fox has had a chance to take each dead bird away in turn. So the impression given to the distraught owner is that the fox has simply killed the lot for fun, when in fact it has been caught mid-scavenge.

TIP 70: *Consider fencing to keep ground predators at bay*

Even in built-up areas, the presence of streetwise, urban foxes, dogs, and coyotes means that few chickens are completely safe from these wily operators. If you want to keep ground predators out using only a physical barrier, then you'll need to build unsightly, permanent fences that are at least 8ft (2.5m) high—they can be impressive climbers. Thankfully, the availability of electrified netting provides a far less intrusive, cheaper, and effective option. Foxes and coyotes can sense the electrical current that's pulsing through the netting and they steer clear, even though the flimsy-looking barrier is little more than 3ft (1m) high.

SECURITY MEASURES

TIP 71: *Get your run fencing right*

As the first and, in most cases, last line of defense against predators, the fence around your poultry enclosure is a very important structure. When keepers are permanently "in residence," and there's perhaps a dog in the household, too, common predators like foxes, raccoons, and coyotes are more wary. In those situations an electrified netting setup might be all that's required. For those leaving the birds on their own more often than not, a permanent and much taller fence may be the more secure option. Remember, it's got to be capable of keeping the birds in, as well as undesirable visitors out. (See also tips 151–160.)

TIP 72: *Permanent fencing provides the ultimate defense*

Everything hinges on the quality of the materials used for permanent fences, and how well they're put up. The hardware cloth must be supported by sturdy posts, with those at the corners being braced and set in concrete for the best results. The mesh must also be well tensioned and, as well as being at least 8ft (2.5m) high, should extend either 20in (50cm) or so down into the earth, or 5ft (1.5 m) outwards from the base of the fence, staked securely to the ground. The object is to deter foxes, raccoons, or weasels from attempting to climb over or dig under to get into the run. Many keepers who go to this sort of expense and effort supplement a fence of this type with runs of single-strand electrified wire at the base, half-way up, and around the top, just to be doubly sure that nothing's going to get inside.

TIP 73: *Electrified netting is effective and flexible*

One of the real beauties of opting for electrified netting is that it's completely flexible, both in terms of the shape of the area enclosed, and because it's so easy to relocate. New sections can be added to enlarge the run size when necessary and these are typically available in different lengths, including 82ft (25m) and 164ft (50m). The ease with which this netting can be moved is a real plus, so it's quick and simple to move your birds on to fresh ground.

TIP 74: *Don't be worried about electrification*

The use of electrified wire may sound a bit scary, but don't let yourself be put off because the advantages really are fantastic. The current involved —which is pulsed down the wires once every second or so—isn't sufficient to be dangerous to humans or pets, although babies and toddlers are obviously best kept away. The "jolt" you'll receive by touching electrified netting isn't pleasant and will certainly make you recoil. Some kits may include a warning sign to hang on the fence.

TIP 75: *Match the energizer to the fence*

One of the key requirements of any electric fence is that the power coursing around the wires is actually sufficient to act as a deterrent to anything that touches it. Electricity loses its potency the further it travels, so longer runs require more oomph. It's important that the power supply— the energizer—is up to the job, so you must buy the right unit for the length of electrified fencing that you intend to power. If you subsequently decide to expand the size of your fenced enclosure, you'll probably need to upgrade the energizer as well, to maintain effective protection for the birds inside.

TIP 76: *Electrified netting must be kept in tip-top shape*

Be under no illusions that electrified poultry netting can be ignored once installed. Although the electrified wires don't extend right to the base of the netting, it only takes a few inches of vegetation growth before leaves and stems can reach the lowest one. When this happens, it creates the possibility for short-circuiting, especially in wet conditions. Electrical "leakage" to earth in this manner will rapidly run down the battery that's powering the system. So it's very important to control grass and weed growth around the base of the netting. Some keepers trim or mow regularly, while others use a weed killer or suppressant membrane to solve the problem. You'll also need to ensure that the netting remains taut. Most setups rely on guy ropes to support the corner posts, thereby tensioning the netting runs. Keep these as tight as possible at all times.

TIP 77: *Power failure; the nightmare scenario*

It's well known that foxes and coyotes have territories that they patrol regularly in their search for food. Animals local to you may well include your chicken run on their daily/nightly route. So, if your electric fence happens to be off when the fox comes a-calling, it'll be into the run in a flash. It's amazing how often keepers report that, the very first time that their electric fencing failed, the fox struck with devastating results.

TIP 78: Protecting birds might require a covered run

Birds of prey can be a real nuisance, especially to those keepers who breed and like to get their youngsters outside and into the sun at the earliest opportunity. Your average hawk or eagle will have no regard for the potential rarity of a breed; all they'll see is a tasty meal tottering about on two legs. Apart from keeping your birds inside, the only other option is to enclose their run with a wire netting roof or clear, corrugated plastic panels.

TIP 79: A secure chicken coop remains a must

Regardless of the level of security you think that you've built into your enclosure fence, it's still essential that you establish a second line of defense; a lockable pop-hole door on the hen coop. The coop itself must be structurally sound too, so that if a predator does manage to get into the run during the night, the birds inside the coop will still be safe. Don't forget the rest of the structure, too. The lid of the nest box and the main access door should be secure as well, and it's vital that the coop itself is screwed together tightly. Loose or rotten paneling presents an open invitation to prowlers, so it should be repaired/replaced as soon as it's spotted.

TIP 80: Don't forget the possibility of theft

The rise in poultry-related crime is fueled by the ease with which stolen birds can be sold, either via auctions or the internet. In some ways, the hobby is falling victim to its own popularity. For this reason, it's best to source your birds from a reputable breeder who has been recommended to you by either a breed club or a trusted fellow keeper. There are plenty of antitheft kits available, ranging from movement-activated lights to full-blown alarm systems that will text your cell phone when there's trouble.

BEDDING

TIP 81: *Bedding can be both a blessing and a curse*

It's all too easy to underestimate the importance of good bedding material inside a chicken coop; it has the potential to have a serious impact on the general health and well-being of the birds. Not only is the basic material choice important, but so is the way it's used and how it's managed. The birds will feel happiest and most settled in their coop and nest boxes if the floors are covered in a decent layer of bedding material that's several inches thick. It helps encourage them to lay, and will also protect those eggs that are produced from being damaged.

TIP 82: *Keep the bedding clean; that's the secret*

To work effectively, poultry coop bedding has got to be clean. It quite naturally becomes soiled during the night as the birds roost. One of its functions is to contain the droppings, of course, but allowing these to build up to any degree is likely to cause problems. Not only does it increase the risk of bacterial development, but its acidic nature can have a seriously detrimental effect on the atmosphere inside the coop. As a producer of ammonia, chicken droppings need to be cleared from the confines of a hen coop before the air in the enclosed space becomes too tainted. Understandably, breathing in this toxic gas isn't good for chickens and can promote all sorts of nasty respiratory conditions among the birds.

TIP 83: *Day-to-day management is the best approach*

Rather than let bedding material become seriously dirty and contaminated before changing it, it's far better to deal with the birds' "deposits" on a daily basis. "Poo picking" takes only a matter of moments every morning—a gloved hand and a bucket are all you need—and is a genuinely good thing to do. It'll greatly increase the overall life of the bedding layer, because the liquid content in the droppings isn't given time to soak in and contaminate the surroundings. This is important for financial as well as welfare reasons. Like everything else, good bedding material is becoming ever more expensive. It's a regularly needed consumable, so anything you can do to extend its working life (and reduce overall consumption) is going to save you money.

TIP 84: *Avoid using hay and straw*

Traditionally, hay and straw were very popular bedding options. Nowadays, though, most experienced keepers agree that wood shavings offer a better and more practical alternative. The problem with straw—especially—is that it can disguise problems. Droppings and moisture fall through the upper layers to lodge deeper down, where they set about their destructive work. The danger then is that they remain unseen by the keeper. Outwardly the bedding layer looks clean and dry, so it tends to get left, untouched, for too long. This gives mold, bacteria, and noxious gases time to develop and start polluting the environment. Of course, it's perfectly acceptable to use straw in a poultry coop if you're a vigilant sort of keeper who will be checking and changing the bedding very frequently.
If not, though, these materials are certainly best avoided.

TIP 85: *Steer clear of sawdust bedding, too*

Another bedding option that has become increasingly popular in recent years is sawdust. In reality, though, this can create serious problems for chickens because of its dust content. With hens being shut inside their coop for at least eight hours every night, the atmosphere within needs to be as clean and fresh as possible. The dust associated with sawdust can cause breathing and eye problems, both of which will be significant triggers for stress among the birds, as well as potentially serious, indirect medical conditions. Even when using wood shavings, it's vital that these are dustfree, and made from soft rather than hard wood. The latter can cause splintering, which is a problem for the birds' feet.

TIP 86: *Good chicken coop design pays dividends*

While you'll undoubtedly pay more for a well-designed hen coop in the first place, the benefits are likely to stand you and your birds in good stead for many years. At a practical level, a coop that features a sensibly large main door, allowing easy access for cleaning, is a real advantage. A door that is too small will prove to be a constant annoyance, especially when the weather turns wet, windy, and cold. A droppings board is another big plus point too. It's essentially a floor-sized panel that sits under the roosts, covered in bedding material. Come cleaning time, the board can be easily slid out, together with its covering of used bedding, and transferred to a nearby wheelbarrow for sweeping clean. A well-designed droppings board will conveniently remove most of the dirty bedding at one time, leaving a minimal amount to be swept out from around the edges of the coop.

TIP 87: *Modern alternatives can work well*

Over the years, chicken keepers have tried all sorts of bedding materials, but some of the more successful alternatives to the traditional options have come from the horse market. You can buy hemp-based bedding that is dust-free and claimed to be more absorbent than straw or wood shavings. It's also suggested that this bedding is more repellent to flies, which is another thing in its favor. However, you can also expect to pay perhaps 30% extra for this type of bedding over shavings, so you might need to make some practical comparisons to get at the real costs of the two options. You can also buy shredded cardboard bedding material that is very absorbent too, a good insulator, and, like the hemp-based option, fully biodegradable (so it's great on the compost pile after use).

TIP 88: *Clean nest boxes mean clean eggs*

One of the biggest advantages of daily "poo picking" in the nest boxes is that it'll help keep your eggshells clean. While not the complete answer to pristine eggs, collecting the regular buildup from the boxes goes a long way to ensuring that your delicious, healthy eggs look as good as they possibly can. It's often the case that one or two of your hens will prefer to roost in the nest boxes overnight, instead of on the roosts, and this is a major contributor to fouling in this part of the coop. But this isn't the only cause of soiled shells. Birds with dirty bottoms will, quite naturally, pass on some of this unpleasantness to the eggs they lay. Ideally, of course, healthy hens shouldn't have this problem, and its presence points to a looseness in their droppings, which, in turn, suggests a digestive condition that may well need veterinary attention.

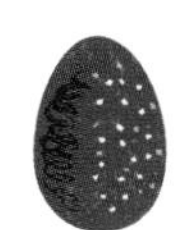
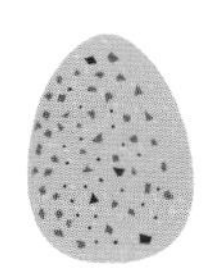

TOOLS

TIP 89: *Keep it simple; it's a low-tech hobby*

One of the beauties of keeping a few hens in the backyard is that virtually anyone can get involved. As long as you have a bit of common sense, a caring attitude and a degree of sensitivity towards the basic needs of creatures kept in captivity, then looking after chickens should be well within your grasp. There's no requirement to invest in a whole range of high-tech equipment to get the most from a small group of birds in a domestic environment. Attention to detail and a methodical approach, coupled with a basic understanding of the husbandry fundamentals, are just about all you'll need to ensure a decent supply of tasty eggs for the kitchen.

TIP 90: *It's sensible to have a basic first-aid kit nearby*

Keep a basic first-aid kit handy, for both you and the birds. As far as you're concerned, just about the worst you're likely to experience is scratches and maybe the odd small cut. You'll probably be dealing with wire fencing, which can have sharp, pointy ends, and may also receive the odd scratch or peck from a feisty bird. Injuries to the birds themselves tend to be pretty minor in most cases and, if things turn serious, then it's usually a job for the vet. However, petroleum jelly (Vaseline®), a good antilouse powder, some antiseptic ointment, a recognized poultry tonic, a chemical wormer, and a vitamin supplement are all items that are bound to prove useful at some point or other during your time with chickens.

TIP 91: *Your hands are the best tools in the box*

It's really important to be involved with your birds, and to handle them on a regular basis. Not only will this enable you to learn a tremendous amount about their day-to-day health, but it'll also help build a valuable level of trust between your birds and you. Adopting a hands-on approach really is the only sure-fire way of keeping tabs on the presence of external parasites, and making meaningful assessments about how the birds are eating and whether or not they're gaining weight. The prominence of the breast bone is a key guide in this respect. Also, the gap between the tips of the two pelvic bones, just below the vent, offers a very useful indicator about whether or not a hen is about to come into lay; more than two fingers wide and things are about to start happening!

TIP 92: *Otherwise, a garden storage box keeps tools tidy*

Keeping everything together in one handy place makes a lot of sense. If you don't want a potentially ugly-looking shed in your backyard, then how about a far less obtrusive garden storage box? You can buy these in all shapes and sizes nowadays, and at prices to suit all budgets. Typically they're made of heavy-duty plastic, which are great, and they have lockable lids or doors, which are even better. Such units can provide convenient and dry storage space for feed sacks, spare fence posts, bedding material, cleaning equipment, rat poison, and your first-aid kit. However, it's unlikely that these structures will be rodent-proof, so putting two or three traps on the floor is a very sensible precaution to take. Whether you're in an urban or a rural environment, there's a seemingly endless supply of these tiny pests that will take any opportunity for a free lunch!

TIP 93: *Gloves are essential, as is a feed scoop*

A decent pair of gardening gloves is key for coop cleaning duties, and the rubberized type are best. Although they make your hands sweat a bit, they're simply the best for "poo picking" as well as digging out soiled bedding from those hard-to-reach nooks and crannies in chicken coops. Another must-have is a simple feed scoop. These can be bought in either plastic or metal, and should be readily available from your local feed store or specialist poultry outlet. They really do make transferring feed from the sack to the feeder easier and less wasteful—the less you spill, the fewer mice, rats, and wild birds you'll attract, which is an important consideration.

TIP 94: *You'll need a wheelbarrow and a dustpan and brush*

A decent lightweight wheelbarrow—if you haven't already got one—is a must. As well as being essential for holding and then transporting used bedding material from the chicken pen to the compost pile, it'll provide a valuable service as a general carrier, helping you transport everything from heavy bags of feed to batteries for the electrified netting. A dustpan and brush is by far the simplest and quickest way of clearing out the used bedding material from your chicken coop. Much, of course, depends on the design of your coop. The best have large-access doors, flat floors, and no sills or thresholds around the edge, so that everything can be swept out with ease.

TIP 95: *Rodents can damage your equipment*

While rats and squirrels might not pose such a threat in your garden storage box, they are certain to be sniffing around the poultry pen in search of food and shelter. The best advice is to go on the offensive from the start with your chosen antirodent measures: it's far better to be maintaining a deterrent at all times so that any local population never gets the chance to establish a colony in or around your poultry run. Rats and squirrels, like foxes, are great opportunists, and will take any chance they can to gain a foothold. They pose a threat to your birds, and can cause damage to your equipment.

TIP 96: *They're seldom alone for long, either*

The trouble with rodents is that, if you see one, there are almost certain to be more in the vicinity. They rarely live alone and breed very quickly, which is why problems can escalate so rapidly. Rats can start breeding at about three months old and, if the conditions are right, will produce up to eight litters a year, each of which will consist of between six and eleven youngsters. So that's a potential for nearly ninety young rats from a single, adult pair, all of which will be weaned after just one month. The conditions under a chicken coop that sits directly on the ground can be just about ideal for this level of reproduction; it will be warm and secure with plenty of food and water within easy reach whenever the rats need it.

TIP 97: *So, track those nocturnal pests*

A common problem that many chicken keepers experience when attempting to deal with rats is failing to appreciate the basics of their behavior. A lack of understanding in this respect will almost certainly limit the effectiveness of any measures taken to control them. Rats are essentially nocturnal creatures, although they will be seen during daylight hours if food is scarce and/or overall numbers are high. They are naturally suspicious of anything new and will rarely cross open spaces, even at night, preferring to keep to the base of walls or along hedges. Traps and bait stations need to be set at these locations when activity is suspected. You can buy fluorescent tracking dust that will reveal their movements when viewed under ultraviolet lighting.

TIP 98: *And maintain traps once or twice a day*

Plenty of people prefer to use rat traps instead of poison; the latter can pose problems in the domestic environment, and also affords no control over where the rat eventually dies (rotting corpses can smell terrible). However, the use of rat traps that catch the animals live can leave you with a further problem: what to do with the trapped rat. Wildlife trapping laws vary from state to state, so check your state's specific laws.

TIP 99: *Use an electric fence tester*

Simple fence testers are a very useful accessory for keepers using battery-powered electrified netting to protect their hens. You can buy these in a number of different forms: some hang on the fence and flash to indicate that all's well, and others have to be manually connected to check the current flow. The basic aim of them all is to make sure that there's sufficient juice being pulsed around the netting by the energizer to provide a useful fox deterrent. Electrified netting is worse than useless if the current flowing through it drops to a trickle due to a nearly dead battery or significant electrical shorting to earth. The latter is commonly caused by vegetation growing up and coming into contact with the electrified wires, and is a particular problem in wet conditions.

TIP 100: *Choose the right battery as a power source*

If you aren't able to power your electrified poultry fencing directly from a main or solar supply, then a 12-volt battery is the natural alternative. However, it's important that when buying this you specify a lithium battery rather than a conventional car battery, as these are designed for the sort of deep cycling that's likely to occur when it's used to power chicken netting. Normal car batteries work best when their state of charge is constantly maintained by the vehicle's alternator. A sequence of a few drainings and rechargings will wear out the unit. Lithium batteries—designed for the camping, boating, and golf cart markets—are manufactured specifically to withstand this sort of operational strain and, consequently, have a much longer working life.

CHOOSING YOUR BIRDS

One of the fascinating aspects of keeping chickens at home is the huge range of breeds that are available to choose from. With more than a hundred established pure breeds, ranging from the essentially plain to the outright bizarre, plus an increasingly impressive selection of hybrids, today's chicken keeper is spoiled as far as breeds to choose from. However, as you'll discover, there's plenty to consider before making your final decision.

BREED BASICS

TIP 101: *There is a difference between pure breeds and hybrids*

One of the most fundamental decisions to be made if you want to start keeping some hens at home is whether to opt for pure breeds or hybrids. In a nutshell, hybrid hens will nearly always produce more eggs than pure breeds—hybrids have been carefully bred by crossing and recrossing the best-performing strains from the traditional laying breeds. The results are feathered "laying machines" capable of producing over 300 eggs in their first laying season. This level of performance is head and shoulders above most pure breeds, and is the main reason why most novice keepers begin with hybrids.

TIP 102: *For character, choose pure breeds*

While not usually as productive as hybrid hens in terms of egg-laying performance, the appeal of the traditional pure breed runs a good deal deeper for most owners. These breeds have a place in history that hybrids simply can't match. The oldest date back literally thousands of years, and most have breed clubs dedicated to their support and promotion. Some have been pivotal in the creation of modern hybrid laying strains, while others have enjoyed complicated and often fascinating histories that keepers love to buy into. So, while most won't give you as many eggs as the hybrid equivalents, they do offer tons of interest as well as being worthy subjects for conservation. With some breeds teetering on the brink of extinction, more keepers are always welcome.

TIP 103: *For convenience, choose hybrids*

In the same way that mongrel dogs tend not to suffer with many of the characteristic failings that can blight the health of pure breeds, hybrid hens are generally great "doers." They simply get on with life, without drama or fuss. Also, having been reared commercially, all should have received effective vaccination against the common poultry problems, making them ringers for a trouble-free laying life. There will always be exceptions, of course, but you'll need to be really unlucky to run into serious health problems with well-sourced hybrids.

TIP 104: *And remember, hybrids and pure breeds need to be housed separately*

One thing to be aware of is that the very thing that normally keeps hybrids healthy and fit can spell disaster for unvaccinated, pure-breed birds. Keepers who move on to pure breeds after having started with hybrids—a common progression—often make the mistake of putting their new birds in the same pen as their old ones. Generally speaking, the pure breeds, because they've been hatched and reared by enthusiasts, won't have been vaccinated (because it's a relatively expensive business for birds produced in small numbers). Therefore, the "live" vaccine being carried by the hybrids can actually infect the unprotected pure breeds, causing disease and even death. Segregation in separate pens is the only answer.

TIP 105: *Think about the temperament of the birds*

For those new to poultry keeping, the last thing you're going to want is birds that are overactive, excitable, and difficult to handle. Those with children will want their birds to have docile characters that encourage involvement from the young ones, helping to promote their interest and curiosity. Characters among the pure breeds do vary significantly, so it's important to match the birds to your expectations. As a general rule, you'll find that the "heavy" breeds are the most calm and docile, while the "light" breeds tend to be more highly strung (see tip 108). Much, however, depends on the amount of input you're prepared to lavish on your birds, and what age they are when you buy them.

TIP 106: *Most hens can be tamed*

A hands-on approach is the key to hens that are easy to handle. If you want your birds to be easy to pick up and confident in your presence, then you need to spend plenty of time with them—i.e., on a daily basis. Chicks that you hatch yourself, and that can be handled from day one, are the easiest to mold into well-behaved and friendly fowl. However, if you buy adult birds, things can be very different. Much depends on how they have been treated up to the point at which you acquire them. Even so, the same rule still applies: if you don't get yourself in among your birds regularly, then you can't expect them to be completely comfortable around you. This, in turn, will make your day-to-day husbandry activities more difficult.

TIP 107: *Know your chicken classifications*

Pure-breed chickens are classified in a number of ways, based on their physical characteristics and feather type. For a start, there are the "soft" and "hard" feather breeds. The hard-feathered types are identified by their close, tight, and sparse feathering, and include breeds such as the Indian Game, Old English Game, Ko Shamo, and Malay. By contrast, the soft-feathered breeds typically present a profusion of soft, luxuriant plumage, and this larger group includes breeds such as the Brahma, Orpington, Rhode Island Red, and Silkie. Then there's a size-related classification that separates breeds into "light" and "heavy" groups. The heavies tend to be more docile in nature, and include favorites such as the Dorking, Sussex, and Wyandotte.

TIP 108: *The "light" breeds can be a handful*

A good number of the "light" breeds—Ancona, Andalusian, Leghorn, Minorca, Sicilian Buttercup, and Spanish—have their roots in the Mediterranean, and their temperaments reflect this. Despite being generally smaller than the "heavy" breeds, the "lights" have a reputation for having more excitable characters (most are considered "flighty") and being generally harder to handle. This, of course, does vary from strain to strain, and much depends on the way in which the birds have been reared and are subsequently treated. One of the big advantages of the "light" breeds is that they tend to lay exceptionally well and will usually outperform most "heavies" in terms of both egg numbers and size.

TIP 109: *Don't disregard the bantam option*

While most pure breeds have a miniature or bantam version available (typically about one-quarter the size of the large fowl counterpart), there's also a small group of miniatures known as "true bantams." These birds have no large-fowl version, and so are classified breeds in their own right. This exclusive group includes the Belgian, Booted, Dutch, Japanese, Cochin, Rosecomb, and Suffolk Chequer. All bantams represent a more affordable and convenient ownership proposition, especially for those with small backyards. Although bantam eggs are inevitably smaller than those produced by large fowl, some breeds lay nearly as well as their big relatives, despite having correspondingly smaller appetites.

TIP 110: *There's no need to pay inflated prices*

Although a good many breeds are now considered extremely rare, this isn't generally reflected in inflated purchase prices. Most chickens remain extremely affordable, and it's only at the rarified end of the exhibition scene where prices can escalate. For the most part, sourcing pure-bred birds via the relevant breed club should unearth good-quality stock that you can buy at very reasonable prices. In the main, most enthusiastic breeders are only too happy for new keepers to be showing an interest in "their" breed, and will do their utmost to supply the best they can to genuine customers. Unfortunately, the same isn't always true at poultry auctions and general livestock sales, when "pure-bred" chickens can change hands at ridiculously inflated prices. Not only is the price paid often higher than it should be, but the birds being bought can be of questionable quality.

PURE BREEDS

TIP 111: *There is a bird type for everyone; just find yours*

With more than 100 readily available pure-breed chickens to choose from, you'll have more than enough choices: from the straightforwardly attractive Australorp to the all-singing, all-dancing Sultan. There are some outlandish options too, including the odd-looking but surprisingly productive Transylvanian Naked Neck, or the long-tailed and exotic Yokohama. Some of the hard-feathered birds from the "Game" breed category—such as the Asil, the Shamo, or the Tuzo—offer powerfully upright and visually striking birds with a beguilingly primitive, almost reptilian air about them. A great many keepers opt for breeds from the most popular, soft-feathered heavy and light categories. Choices such as the Orpington, Leghorn, Sussex, and Silkie continue to prove popular among those who support breeds that so many breeders have worked so hard to preserve over the generations.

TIP 112: *They're not all great layers nowadays*

If you want a good supply of eggs for the kitchen, and are set on keeping pure breeds, then you'll need to take care when choosing and buying your birds. A good proportion of the breeds don't lay as well as they once did, thanks, primarily, to the efforts of those interested in showing who strove to make these birds more attractive and successful on the exhibition bench. This typically involved careful selection to promote the production of greater size and increased feathering, both of which can come at the expense of egg production. So, when buying, it's important to establish with the seller whether the birds are from an exhibition strain or a more productive, utility one. Unfortunately, not all sellers will answer this sort of question honestly nowadays, which is another reason why it's always best to source your supplier via a recommendation you can trust.

TIP 113: *Some breeds can be very high maintenance*

It's important to do some research into the breed you're planning to buy before finally taking the plunge. You may imagine that, because they're all chickens, each breed is going to have similar requirements on a day-to-day level. Well, in fact, this isn't the case, and there are marked differences in "user-friendliness." Birds with feathered legs and/or head crests, such as the Brahma and Polish, can pose additional challenges for keepers, as the extra feathers can both mask and cause problems. For this reason it's perhaps best to regard breeds like this as something to aspire to, if you're a novice. It's generally far better to gain your early experience with a more straightforward breed, one that's simple to manage in all respects.

TIP 114: *Match a breed type to your needs*

Given that there are such differences between breeds, it makes a lot of sense to think carefully about matching the one you choose to your actual needs. The last thing you want is to be disappointed a few months down the line, as the birds you've bought fail to meet your expectations, or simply don't suit the environment in which you're asking them to live. Once again, it comes down to doing the research before you get started. Don't simply get overtaken by the look of a breed; do all you can to understand its individual needs and likely behavior before you get involved. Get in touch with the relevant breed club, visit a show or two and talk to experienced breeders. Most will be only too happy to share their knowledge with you, and offer helpful, practical advice that could very well save you from making a mistake.

TIP 115: *Game birds can attract unwanted attention*

The sad reality is that cock fighting still exists today. Although outlawed in all 50 states, this barbaric practice continues on an "underground" basis, so there's a constant demand for Game birds. You may want to take this into account if you're considering keeping Game birds. For a start it can be hard to get stock in the first place, as genuine enthusiasts and breeders are naturally suspicious of new customers appearing out of the blue and, if you get birds, then your security measures will need to be up to the job.

TIP 116: *Always breed for the best results*

Most people who start keeping pure breeds find that, before too long, they want to get involved with breeding their own birds. The temptation to start producing chicks at home is simply too hard to resist! However, for the sake of the birds you're breeding—and the breed itself—it's important to make the effort to breed properly. If you're serious about what you're doing, then you should aim to hatch chicks from decent parental pairings. Put your best-quality birds together and take a pride in the results achieved. Try to avoid crossing in a haphazard manner, as all this does is weaken whatever bloodline you have. It's far better to take a responsible approach and, if needs be, borrow a good male bird from a friend or fellow club member to ensure that the birds produced are of the best possible quality.

TIP 117: *Careful selection is vital for breed survival*

The conservation angle is a significant one, as far as pure-breed poultry is concerned. However, it's equally important to appreciate that, for the good of the breed, quality should be at the core of any breeding program you start. Just because a bird looks a bit like a New Hampshire Red—or is sold as such at a poultry sale or auction—doesn't mean that it actually is one. People may pull all sorts of tricks these days, now that they see there's a bit of money to be made selling chickens. It's vital that you know a bit about the parentage of the birds you want to breed with. Some of the rarest of our pure breeds now exist in worryingly low numbers, so careful, well-considered help is needed with these birds, not indiscriminate cross-breeding.

TIP 118: *Pure breeds enjoy a longer working life*

One of the most significant advantages of pure-breed chickens over modern hybrid layers is that, although they don't produce as many eggs early on, they do continue laying for much longer. Although a typical hybrid hen will lay at least 300 eggs in her first season, the number in the second year will fall away quite dramatically, to perhaps 200 or so. Thereafter, totals will tumble even more. This is due to the fact that hybrid hens are working birds, and have been carefully developed to lay fantastically well for around 18 months. Beyond this point, though, it becomes uneconomical to keep them on a commercial scale, so they are simply replaced by fresh, young birds. In contrast, a healthy, well-kept pure-breed utility hen can continue laying good numbers of eggs for four or five years.

TIP 119: *Demand for roosters is usually limited*

One thing to remember, if you decide to start breeding, is that about half the chicks you hatch will be males, which could pose a problem. For those of you living in urban areas—or even in the countryside, but with neighbors close by—the early-morning crowing of roosters is likely to come as something of a rude awakening. Regardless of how many free eggs you might offer as compensation for what many consider an antisocial disturbance, it's likely that tempers will become frayed and relationships will quickly become far from neighborly. For these reasons, lots of people choose not to keep roosters and this, as far as breeders are concerned, can throw up a bit of a disposal problem. So, unless you feel able to kill any surplus males yourself (or have an experienced friend who will do it for you), getting rid of those you don't want will probably prove difficult.

TIP 120: *Rare breeds are always in need of more support*

With the total number of surviving birds among the rarest breeds numbering just a few hundred, there's no question that the strugglers need help. Breed clubs are at the core of the conservation effort, but even some of these are struggling to stay afloat. What's more, some breeds don't even have a club to support them; there simply isn't the number of keepers out there to warrant one. Thankfully, these fading breeds are taken under the protective umbrella of the Society for the Preservation of Poultry Antiquities (SPPA) and the Heritage Poultry Conservancy (for website details see page 288), where a registrar exists to coordinate the efforts of the few breeders who remain in each case. Typical of those breeds currently being assisted by the SPPA are the Golden-Laced Polish, Sicilian Buttercups, Salmon Faverolles, Black Copper Marans, and Silver Phoenix.

HYBRID LAYERS

TIP 121: *First things first, hybrids are friendly fowl*

If there's one thing you can be sure about with hybrid hens, it's that they'll be delightful birds to be around. One of the primary objectives of their careful and exhaustive selection process was to create hens with docile, nonaggressive characters. This was essential as far as the commercial operators were concerned: the last thing they wanted were sheds full of thousands of intensively cooped hens all fighting each other instead of laying eggs. Domestic keepers are now able to benefit from this development work and have the option to enjoy hens that—if kept well—will be easy to manage and a pleasure to handle.

TIP 122: *And make just about the ideal "starter bird"*

Rather in the same way that a mutt offers the best and least troublesome option for the would-be dog keeper, so the humble hybrid hen can represent the perfect entry point for those wanting to get involved with chicken-keeping. Pure-breed hens, while providing great history and superb variety, can be prone to problems. Just as some breeds of dog are known sufferers of breathing difficulties or back problems due to the amount of in-breeding that's gone on in the past, so certain breeds of chicken are recognized as having genetic weaknesses and being susceptible to specific conditions or disease. Hybrid hens, however, don't suffer in this way, and all but guarantee their keepers a trouble-free and straightforward life.

TIP 123: *Buying birds should be straightforward*

Sourcing hybrid laying hens shouldn't be a problem these days, assuming, of course, that you seek out a decent supplier of genuine birds. As always, do some research first, and take advice and recommendations from those in the know. The growing popularity of poultry keeping means, unfortunately, that there's an increasing number of individuals setting themselves up as "specialist breeders," so never be afraid to ask questions. Poultry magazines and internet sites will contain plenty of ads and links promoting hybrid hen suppliers, but it's important to make sure that those you buy from are providing properly bred stock and not home-made crosses with a rag-tag genetic background. If you detect any sort of doubt in response to your inquiries about the background of the birds being offered, then simply walk away and find another supplier. Genuine outlets will know precisely what their birds are and where they came from.

TIP 124: *There's plenty of plumage color choice nowadays*

It wasn't too long ago that those wanting to keep hybrid hens only had a choice between brown- or white-feathered birds. All that's changed now. Today there's an ever-increasing range of plumage colors available, so it's no longer the case that hybrids have to look boring. Recognizing the potential for something a little different in the domestic market, breeders have been hard at work creating birds that they feel offer keepers the best of both worlds: great egg production and looks to match. So, in visual terms, it's no longer the case that hybrid hens are the poor relations of the pure breeds.

TIP 125: *Want colored shells? Hybrids can deliver*

With the blossoming of plumage color options among hybrids has come the possibility of pretty eggshell colors too. Not so long ago, all you could expect from these birds was either a white or a tinted (pale brown) egg; the colors matched the plumage of the birds in those days. Now, though, things are a good deal more exciting. Hybrids with the Araucana and the Marans in their genetic makeup have been produced, resulting in egg-shell colors that include pale greens and blues, a range of browns, including a dark, chocolate-like version, plus speckled. So today, not only is it easy to put together a really attractive-looking, backyard flock of hybrid hens, it's also possible to produce a beautiful array of colored eggs that look fantastic in a bowl on the kitchen counter.

TIP 126: *Rescuing an exbattery hen can reap great rewards*

Adopting exbattery hens is a relatively new phenomenon, but one that's caught the imagination of many thousands of compassionate British keepers. The British Hen Welfare Trust (BHWT, formerly the Battery Hen Welfare Trust) is the leading charity associated with the rehoming of commercial laying hens, and has overseen the transfer of more than 250,000 birds out of intensive, commercial farm environments and into gentle retirement with caring new owners. The United States does not yet have an equivalent organization, but some poultry farms may give away their layers when they are past their prime (generally at about eighteen months of age). If you choose to go this route, select alert, healthy, agile birds. Many battery-cage hens are so spent from their living situation that they are very susceptible to disease.

TIP 127: *Exbattery birds can make gentle garden companions*

Although there's no question that a well-adjusted, exbattery hen can become a gentle and appreciative feathered friend, it's important to realize that not all hens with a commercial background turn out this way. Part of the problem is that retirement in a rural location comes as a tremendous shock to these birds. Having spent the first 18 months of their lives cooped up in a warm, safe, and relatively dimly lit environment—with food and water available 24/7—the transition to the great outdoors, with its unfamiliar noises, big skies, sun, wind, and rain, represents a major upheaval. Couple this with the tendency for many of these birds to be decidedly sparse in the plumage department, and the acclimatization process won't always be a smooth and uneventful one.

TIP 128: Allow rescued hens plenty of recovery time

You can't expect freshly "rescued" exbattery hens to adapt to their new surroundings overnight. They'll more than likely arrive with you somewhat traumatized and, consequently, will need time to readjust and gather their strength. Stress is an extremely draining condition for chickens, so it's vital that you do everything possible to minimize stress during the transitional period. Keep the birds shut in their coop for the first 24 hours, with a supply of fresh water and good-quality feed. Then, when you open the pop-hole, let the hens venture outside in their own time: never chase them out before they're ready. Keep noise and other disturbances to a minimum for the first few days. Excitable children and other pets should be kept at a safe distance until the birds are moving in and out of the chicken coop in a normal and confident manner. The whole readjustment process can take a number of weeks.

TIP 129: Exbattery birds tend not to lay so well

The fact that intensively farmed laying hens lay so well during the first eighteen months of their lives means that they shouldn't be relied upon to be so productive thereafter. It really will be a bit of a variable feast, as far as laying performance is concerned. Some birds will produce a decent number of eggs for several more years, while others may hardly lay at all. None is likely to match the 320-plus a season totals of its early life. Most people "adopt" these birds more as an act of kindness than as a way of sourcing a good supply of eggs for the kitchen. Rehoming ex-batts really is an exercise that's more about the issues surrounding the act than what can be gained from it.

TIP 130: *You can always take your involvement further*

No organizations currently exist in the United States for rehoming battery-cage hens. If you are interested in assisting with the process, you might want to approach poultry farms or learning farms, such as Prairie Crossing in Grayslake, IL, which periodically rehomes it's laying flock. You may want to become a "broker," informing people that hens are available for adoption.

THE BANTAM OPTION

TIP 131: *Bantams can be a genuinely practical choice*

There's no question that bantams make a lot of sense, even on economic ground alone. If you're working on a tight budget, then you're unlikely to find birds that will be more economical to feed and require less of a financial outlay (in terms of chicken coop size and run dimensions). What's more, the feed-to-egg conversion rate that most bantams display is excellent, as, in many cases, is the size of the eggs produced. Leghorn bantams, for example, are great layers of good-sized eggs. Obviously, they are smaller than those produced by the large-fowl hens, but not by that much, and the numbers laid can be pretty much on par, too. The general exception to all this are the true bantams, which tend to lay much smaller eggs and in much fewer numbers. However, in their defense, these birds are never kept for their productivity.

TIP 132: *If you are interested in showing, start with a bantam*

Because they're smaller, there's less work involved in preparing a bantam for an exhibition; it's a straightforward matter of scale. This tends to make them a rather less daunting prospect for the show novice, which can be a big advantage. What's more, there's no question of bantams being regarded as the poor relations of their large fowl counterparts. So, a good and well-prepared bird will stand just as good a chance of winning as any other. However, this isn't to say that showing bantams is any less competitive, or that it represents an easy option, compared to exhibiting large fowl. There isn't quite so much work involved, but the standard of excellence required for success at the top level is every bit as demanding. (See also chapter 10.)

TIP 133: *Bantams are more economical than large fowl*

The general manageability of bantams is one of their most attractive attributes, to new and experienced keepers alike. The fact that they are generally cheaper to buy is increasingly relevant today, as is the reduced space requirement when compared to keeping large fowl. These factors count greatly in their favor, and are major contributors to their great popularity. But aesthetics play a significant part as well. In some cases, the type and beauty of the bantam version can be superior to that of the corresponding large fowl; they can represent the very essence of a breed. Many people simply prefer the look of the miniature, and the economic advantages merely represent an added bonus.

TIP 134: *True bantams can be pint-sized perfection*

For those of you with a hankering for visually stunning chickens then, by common consent, you need look no further than the Sebright. This true bantam, available in either silver- or gold-laced forms, is the absolute epitome of all that's good about miniature fowl. Good examples are neat, stylish, sharply marked, active, alert and generally extremely beautiful. The Rosecomb is another example of a wonderfully attractive bantam presenting tremendous visual appeal, albeit without the striking feather lacing possessed by the Sebright. However, if you fancy something a little more feathery, then the Belgian Bearded bantam—in Bearded d'Uccle form—offers fluffy profusion under the beak and on the legs and feet. It's the Cochin, however, that is perhaps the feather lover's favorite; plumage is so profuse that the short legs are completely hidden.

TIP 135: *Small bantams can be more vulnerable to extremes*

One potential downside of the very smallest bantams is that they're not generally as hardy as larger birds. Prolonged periods of cold or very wet weather can cause problems, with chilling being a real concern. Therefore, it's important that the conditions in which these birds are kept are as good as they can be. Covered runs are perhaps the ideal approach during the winter months, so that feathered feet and legs can be kept as clean and dry as possible. Some keepers opt to keep their small bantams shut in during the coldest months but, if you're contemplating this, it's vital that the accommodation is both large enough to ensure that the birds aren't over crowded, and well ventilated without being drafty. Sensibly converted garden sheds are a popular and relatively cheap option for over-wintering potentially vulnerable bantams.

TIP 136: *Young children often feel happier with bantams*

It's great to encourage children of all ages to become actively involved with domestic poultry; there are so many valuable lessons to be learned about husbandry, life cycles, food production, and welfare issues. However, if you happen to be keeping one of the larger breeds, such as the Brahma, Orpington or Plymouth Rock, getting close to and/or handling birds like this is likely to be a daunting if not downright scary prospect for younger family members. Not only can the sheer size and weight of these breeds make holding them properly very difficult for young arms, but the likelihood of an uncomfortable or insecure bird lashing out with beak or claw in panic is greatly increased. With the right bantams, none of these problems should arise, and they and the children should be good buddies right from the get-go.

UTILITY FOWL

TIP 137: *The utility fowl is an all-round performer*

One of the great attractions of keeping chickens at home is their amazing versatility. They can be pampered pets, egg producers, exhibition specials, reared for the freezer, or form the basis of a business. In addition, of course, they can be any combination of the above that you like. One particularly useful option is the so-called utility fowl; pure-breed birds, which are recognized as both decent producers of eggs and meat. Established, traditional breeds such as the Sussex and Barred Rock can work really well in this respect, as long as you get proper utility strains as opposed to the exhibition types, which won't lay nearly as well. Keepers must be prepared to kill their birds, of course, and this does understandably pose problems for some, especially "pet keepers," many of whom couldn't imagine anything worse.

TIP 138: *Don't go for the "fast food" option*

Modern, broiler-type chickens are easy to source nowadays—even in relatively small numbers—and rearing them shouldn't raise any significant problems for the more-experienced keeper. The one thing to be careful of is feed intake. In the commercial environment, these birds are fed a very carefully controlled diet to ensure that they gain weight at the fastest possible rate. The birds will eat voraciously and, if given too much food in the domestic environment, will tend to put on too much weight too quickly. This can have serious (potentially fatal) health consequences, and so access to feed will need to be properly regulated. Offering a free-choice supply of pellets, as many laying hen keepers typically do, is likely to induce undesirably rapid weight gain, leg problems, and maybe even heart failure.

TIP 139: *Traditional breeds are great for the table*

Of course, there's no real need to rear modern, hybrid meat birds if you're simply after a supply of roasters for the family table. Traditional pure-breed males, such as the Orpington, Sussex, Rhode Island Red, and Buckeye, can be grown on to produce superb, flavorful eating. Many argue that these breeds actually produce meat of a finer flavor and texture than the modern alternatives, although it's unlikely that you'll ever achieve the same killing weight with the traditional breeds. They'll always be slower to mature, requiring more food and care as well. So, if you're breeding stock to sell or show, and usually find yourself having to cull unwanted males, why not raise a few on for the table? The males of most traditional breeds can be eaten, although the color and taste of the meat will vary. Age is a key factor too. Let most birds get much more than a year old, and you risk the meat starting to get tough.

TIP 140: *Choose your meat bird supplier with care*

Commercial-type meat birds are commonly bought as day-old chicks, but you'll need a heated brooder unit if you take this approach. Alternatively, you can buy the birds at five weeks old, when they're just off-heat. The popular Ross/Cobb-type birds can be ready for processing from eight weeks onwards. If you want a slower-growing option, then a Freedom Ranger may be better. These birds were developed in France in the 1960s and can be slaughtered at twelve weeks. There are a handful of commercial breeders, so most should be happy to deal with enthusiast-type keepers; as the market for domestically reared table birds grows, so these specialist suppliers are starting to appreciate that there's money to be made. A quick search on the internet will turn up the various options.

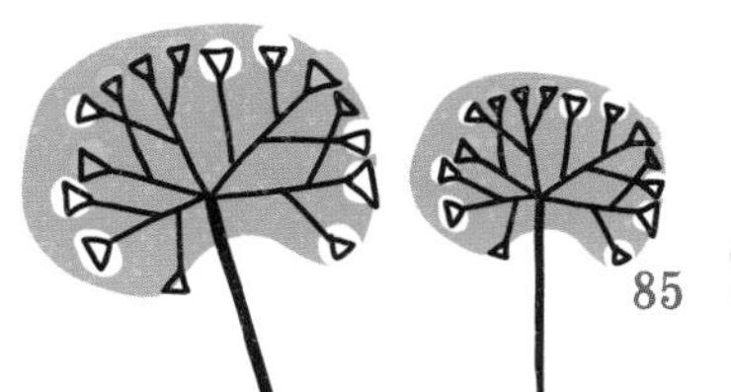

TIP 141: *Focus on quality, not quantity, for taste*

In the same way that home-produced eggs taste much better than those bought from the supermarket, roasting your own meat birds will add a new dimension to your family's consumption of chicken. The meat from birds that have been grown slowly in a relaxed, free-range environment has a taste and texture of its own. Pure breeds (even the utility strains) are very unlikely to ever match the weight of a commercially produced, oven-ready broiler.

TIP 142: *Take confidence in what you're eating*

Processed food can be awash with additives and preservatives, which many take exception to and in recent years, questions have been asked about the excessive use of antibiotics, medications, and other additives. While the authorities have introduced much tighter controls on US producers in this respect, some consumers still have their doubts. Growing your own birds at home insulates you from risk, given that you have controlled what they've eaten and any medicine administered.

TIP 143: *Become comfortable with home dispatch*

Killing poultry (dispatching) is never a happy task; it's upsetting, and will always be so if you value your birds. There's bound to be a time when you're going to need to put a bird out of its misery, be it caused by illness, disease, the aftermath of a fox attack, or some other injury, and a vet won't always be available. It's your duty to minimize suffering. It's important that the method you choose for slaughter is a humane one (see tips 437–442). Commonly used methods are neck dislocation or euthansia. Check with your local laws regaring slaughtering animals on your property, especially if you live in an urban setting.

TIP 144: *Take the DIY approach to feather plucking*

Once you've killed a bird, it'll need to be plucked, and there are three main options at this stage. You can either wet- or dry-pluck it using a machine, or you can do things the old-fashioned way, and remove all the feathers by hand. Wet-plucking involves first dunking the dead bird in hot water, at a scalding temperature of 149–158°F (65–70°C) for about half a minute. This loosens the feathers, enabling them to be removed easily by the rotating rubber fingers on a wet-plucking machine. It's quite a gentle pulling action that minimizes the risk of tearing the skin. This is important for overall presentation of the finished bird. However, you don't have to buy an expensive wet-plucking machine—the feathers can be pulled out by hand, either while the bird is still warm after killing, or after it's been dunked.

TIP 145: *Dry-plucking machines are quick but expensive*

If you've got $2,000 or so burning a hole in your pocket, and more than a handful of birds to pluck, then a dry-plucking machine could be the answer. Although expensive and very noisy to use (ear plugs are advisable), these efficient machines make short work of neatly plucking chickens. Rotating metal discs pinch and pull the feathers out as the bird is rotated by hand in front of them. There's a bit of a knack to the technique; the discs will nip at and tear the skin if the bird is held too close. The pulled feathers are sucked away and piped into a collecting bag for convenience.

ALTERNATIVE CHOICES

TIP 146: *Avoid rearing unwanted male birds*

For anyone using an incubator or broody hen to hatch chicks at home, one potential problem is the number of male birds produced. Crowing roosters cause problems for keepers with close neighbors, and it's often difficult to sex young birds until the obvious differences (comb size, tail feathers) start to become apparent. This means that breeders are forced to pay for the rearing of birds they may not want. However, the aptly named autosexing or sexlink breeds solve this problem—and save the cost of wasted feed—because obvious differences in down color indicate the chick's sex, straight out of the shell. Well-known options include the autosexing breeds Cambars and Jaerhons and Black sexlinks. The practicalities of sexlinkage in chickens were established by Professor Punnett and Mr. Pease, working in Cambridge, England with the Gold Campine and Barred Rock breeds during the late 1920s.

TIP 147: *Sultans take unlimited time and patience*

There's a rare breed called the Sultan, which, if you're looking for one bird that features more or less all the "bells and whistles" you could possibly hope for, will surely fit the bill. With its ancestral roots in Turkey, this snow-white bird boasts head crest, twin-spiked comb, beard, facial muffs, vulture hocks, feathered legs, and five toes on each foot! No other breed offers such a full set of poultry features and, for this reason, it's an extremely challenging bird to breed and show. Getting everything right can be a real struggle, which is probably one of the reasons why the Sultan remains such a rarity. It does have its dedicated followers, of course, but they are rare birds.

TIP 148: *For something completely different, go for a Silkie*

Some breeds possess a completely unique appearance and therein lies their appeal. The Silkie's obvious feature is its fluffy, hair-like feathering. Good examples of these birds—with short, feathered legs—appear as eye-catching balls of fluff, and their heads are topped with a powder-puff-like feathered crest. Other unusual features of this Asian breed is that it is one of the exclusive group of five-toed birds, boasts black skin, a dark purple or black comb, and its wattles and ear lobes can be bright turquoise.

TIP 149: *There is charm to be found with the Salmon Faverolle*

Like the Beared Belgian d'Uccle bantam, and the Sicilian Buttercup, the Salmon Faverolle is a breed whose name has the sound of far-away places. This lovely dual-purpose bird has an upright carriage and a striking beard and muffs. They were bred in the French village of Faverolles from a mixture of genes including Brahma and Cochin as well as local French varieties. These gentle, quiet, sweet birds are good with children. They are also a good cold-weather fowl due to their small combs.

TIP 150: *Asian birds have something to crow about*

If noise isn't an issue for you, then you might fancy one of the breeds of chicken from Asia that have been specially developed for their long crowing—up to twenty seconds long! Primarily from Japan, there are also breeds from Turkey, Russia, and Germany. The Japanese breeds have great names like Koeyoshi, Kurokashiwa, and Tomaru, but they are extremely rare in the US. However, they do turn up at some of the bigger poultry shows.

BASIC RUN SETUP

Thankfully, keeping chickens in your backyard isn't particularly complicated. It's a hobby that's both rewarding and productive, and yet doesn't require a highly technical approach or vast amounts of specialist knowledge. However, there are some important aspects that you must be sure to get right from the get-go. Much is based on straightforward common sense and, as long as you're organized and able to follow a few simple rules, then an easy and pleasant experience with your new feathered friends should be assured.

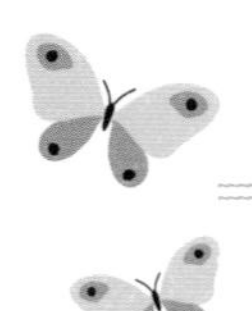

FENCING CONSIDERATIONS

TIP 151: *Understand the purpose of poultry fencing*

Poultry fencing needs to serve two important functions at the same time: as well as keeping the birds contained within the enclosure, it also needs to keep predators out. These days, predators of all sorts are happy to strike, even in broad daylight, so you need to be confident about your defenses.

TIP 152: *You don't have to build Fort Knox*

"Permanent fencing" is supported by wooden posts set in concrete that uses thick-gauge wire stock fencing tightly strung between the posts and buried at least 20in (50cm) into the ground at the base, and extends upwards to a height of about 8ft (2.5m). However, most beginners should be able to create a secure chicken enclosure using electrified netting. Start small and save the big investment for a time when you know you want to begin taking things really seriously.

TIP 153: *Choose the color of electrified netting wisely*

Poultry netting is available in different colors— including green, black, orange, or yellow. If your poultry are going to be tucked away out of sight of the coop, then brightly colored netting won't really be an issue. In fact, it can be an advantage from a visibility point of view, as it's much easier to spot sagging areas and other problems from a distance. However, if chickens are being corralled closer to the coop and in full view of "public areas," go for the subtlety of darker, natural coloring.

TIP 154: *A permanent fence means a low current*

The general advice for fencing types must be that smaller enclosures are more suited to the flexibility and convenience of electrified netting, whereas larger areas are best protected by a permanent fence setup. It's a matter of practicality: the longer the run of electrified netting you have, the more chances there are for things to go wrong (short circuits, sagging, etc.). Keeping a useful amount of current pulsing around a long run of electrified netting (656ft / 200m or more) requires a hefty energizer and efficient connections everywhere. Permanent fences are easier to "protect" electrically, as they will require only two or three single strands of electrified wire, which places less of a load on the supply system. The single strands—if fitted correctly—will also be far more efficient, in terms of insulation.

TIP 155: *Remember, the wily fox needs only one weak link*

Foxes are creatures of habit and "work" their territories with patience and regularity. So, it's vital that the wily fox never gets a single chance to strike. The one time that you forget to check the power level of your battery, or get home late from work and can't be bothered to shut the hens up because it's cold and wet, is when trouble is likely to occur.

TIP 156: *Don't forget to dig it in*

Because permanent fences aren't electrified in their entirety, but are protected by strategically placed single strands of electrified wire, they can present a digging risk. While a fox will certainly think twice about attempting to climb a fence that's protected in this way, it will definitely investigate digging underneath the fence as an easier option. For this reason, it's essential that the fence is extended into the ground, either down vertically or at right angles, away from the base.

TIP 157: *Always stake and support netting carefully*

An electrified netting fence is only ever going to be any good if it's properly erected. With these typically being little more than 3ft (1m) tall, any reduction in the current flowing will encourage foxes in particular to try to jump over. Most kits you buy will include thicker corner posts, designed to be supported by thin guy "ropes." Arrange these with care to create the necessary tension in the netting, or the fence will sag and power-sapping short circuits will quickly lead to the loss of all deterrent value. Also, don't forget to use the plastic stakes—these should be provided in any standard kit—for securing the base of the netting. This will be especially important if you're setting it on bumpy ground. Check carefully around the base of the fence, too; it's easy to miss dips in the ground when they are hidden in grass. Hens may use gaps at the base to escape; predators may use them to get in.

TIP 158: *Make sure everybody knows the fence is electric*

Good electric-netting fence kits will also be supplied with brightly colored plastic warning signs; these are designed to be hung on the fence and to leave visitors in no doubt that the fence is carrying a current. While it's unlikely that any of these fences will be carrying the sort of current that could do any serious damage to anyone who touches them, the shock they give isn't pleasant, and can be quite traumatic for younger children. If you're working to win over your doubting neighbors and, during a visit intended to demonstrate how great your hens are, little Johnny gets a jolt from the fence, you need to be able to point to the prominent warning sign.

TIP 159: *Don't worry, pet dogs and cats soon learn*

Some new chicken keepers worry about how their other pets will react to the presence of electric fencing in the yard. As a rule of thumb, cats seem to take little, or no, notice of hens, while dogs can be a bit clumsier but soon learn from their mistakes—after a few shocking experiences, dogs are likely to become sensibly wary of an electric fence and the chicken netting.

TIP 160: *Use the sun's power to bolster the fence*

If, like most keepers using electrified netting, you're powering the energizer with a 12-volt lithium battery, then the addition of a small solar panel will usefully lengthen the battery's working life between charges. Keeping the battery's output up to snuff is half of the battle when it comes to maintaining an effective electrical deterrent. So, if you're not quite as conscientious with a battery tester as you should be, a solar panel that utilizes the sun's rays to trickle-charge the battery whenever the weather is bright enough is a great bit of extra insurance to have. From experience, one of these panels will extend a battery's working life by five or six months between proper, full charges.

BEFORE YOU BUY THE HOUSING

TIP 161: *Never skimp on the chicken coop*

It's difficult to emphasize strongly enough how important good chicken coop design is to the overall health and welfare of the birds that live inside that unit. With the coop usually being the single largest investment that most newcomers to poultry keeping make, it can be tempting to cut a few corners here and there. Like many temptations, though, this one needs to be resisted. Inexperienced keepers are very much in the hands of the coop seller when it comes to settling on the right unit for their needs. As with so many other aspects of this hobby, experience counts for a great deal when it comes to chicken coop design. So do your research and pay a fair price for a well-planned unit.

TIP 162: *Beware of apparently cheap imports*

Tempting though it may be to pick up a brand new chicken coop at a rock-bottom, bargain price, what you've always got to remember is that these units are cheap for a reason: it's just about impossible to make a good-quality, six-bird chicken coop out of decent wood for $400, and to get it shipped in from Asia. The dangers with these coops are that, although everything tends to look hunky-dory on the face of it, the fit and finish will generally be inferior and the wood used will almost certainly be thinner than you'd wish for. The consequences of these cost-cutting measures only really start to become apparent with time. Exposure to typical winter weather is likely to cause thin wood to start warping, leaks to develop, and drafts to be created—all bad news for the hens inside.

TIP 163: *Ark-style housing can be limited*

Although attractive to look at, many small versions of ark-style coops should house no more than 2-3 hens. Perhaps this has been done to make these units more portable in the domestic environment (they need to be moved on to fresh ground on a daily basis if the birds are not let out), or perhaps it's simply a measure aimed at reducing the cost of manufacture. Either way, the consequence of the downsizing is that headroom for the birds inside can be severely limited. Some examples are suitable for bantams only, as large fowl would only be able to stand comfortably upright in the very center of the apex, at the highest point. This is fine, of course, as long as these units are sold with this caveat. Birds with inadequate headroom in this sort of coop, with its wired panels, are likely to suffer damage to their combs, which will bleed, potentially triggering pecking damage from the other birds.

TIP 164: *Make sure the coop is big enough*

It's always sensible to buy a chicken coop that's just that little bit larger than you need. Roosting birds will always benefit from more rather than less space. It's also important that the atmosphere inside the coop remains fresh and airy and, apart from the basic design, the number of birds inside at night is a key factor in this respect. Another reason to "go large" is because, in most cases, people who get the chicken-keeping bug will start buying or breeding extra birds sooner rather than later. Having that extra accommodation capacity in the chicken coop is always handy. It'll save you having to upgrade your hen coop quite as quickly as you might otherwise have to do. Never forget that overcrowding is a real no-no. It quickly leads to all sorts of health and welfare issues. Undesirable behavioral vices like feather picking, egg eating, and even cannibalism can all become difficult-to-deal-with realities if roosting space is too tight.

TIP 165: *Always ask about the wood quality*

When you're buying a chicken coop, it can be difficult to assess the quality of what you're looking at, especially if there's a glib sales assistant on hand, confusing you with jargon-peppered banter. Using a well-established hen coop builder should minimize the baloney, but there's still no harm in asking a few questions yourself, especially about the wood that's been used. Thickness and treatment are both crucial factors; always opt for properly pressure-treated wood over wood that's been painted with preservative. It's also vital that it be thoroughly dry before being used; damp wood will move as it dries out, opening up gaps and potentially letting in rain. Cut ends are best treated too, to help prevent rotting and splitting as the wood ages. The thicker the timber used, the better. Regard ¾ in (19mm) timber as the minimum for a long-lasting structure. If what you're looking at falls short in any respect, move on to find the right one.

TIP 166: *A pull-out droppings board is always handy*

A removable droppings board is, as the name suggests, a section of flat board that slides across the floor of a chicken coop and is used to catch the droppings produced by the birds as they roost (always more than you might imagine!). Some coops have them, others don't. But if you get the chance, then go for it. While not sounding dreadfully exciting, these simple boards can prove to be an absolute blessing on a practical, day-to-day level. The ability to simply slide out the majority of the hens' production, plus most of the dirty bedding from the floor, too, in one go, is a real plus. Getting rid of so much of it at once usually makes sweeping out what's left quick and easy.

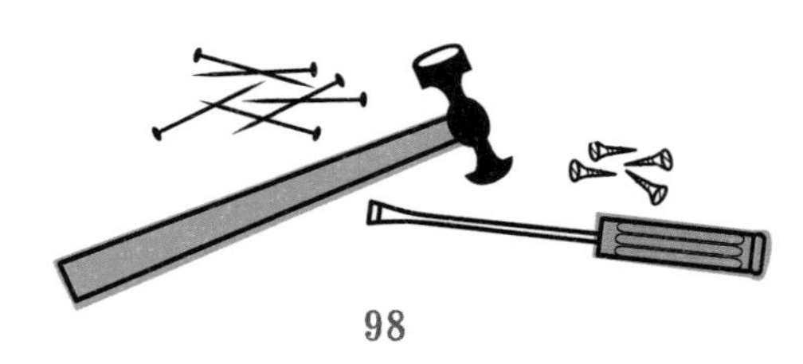

TIP 167: *Think about cleaning out in the rain*

There's always a tendency when thinking about life with chickens, and how great it's going to be to picture things under the golden glow of the summer sun—collecting fresh eggs on balmy June mornings, refilling the waterers on scorching Sunday mornings before settling down to read the newspaper. The reality, of course, is usually different. For a start, it may be raining. There's mud on the ground, your boots will be leaking and you need to clean the chicken coop out again, even though you only did it three days ago. So focus on this less attractive image when choosing your new hen coop. How high is the coop off the ground? Will you be able to clean it out without having to kneel in the mud? How big is the main door. Large enough to get your top half inside, or just big enough for two arms up to the elbows? These things are important, as you'll come to appreciate with experience.

TIP 168: *Don't reject the plastic hen coop option*

Most people would like a traditional-looking wooden chicken coop, presumably because they sit in the yard setting nicely and, well, hen coops have always been made that way. Both these reasons are true, of course, but there is a slowly growing number of people who are learning to appreciate the many practical benefits that plastic chicken coops can offer. For a start, they're often made from recycled plastic, which is great for the environment and saves a tree or two. Also, not being made from countless pieces of wood all screwed and slotted together means they won't be a pleasure palace for the chicken keeper's arch-enemy, the red mites. Combine these important points with an almost limitless and maintenance-free service life, simple repair potential, and the ultimate in easy-to-clean simplicity, and you have a pretty compelling argument. But, they don't look as traditional.

TIP 169: *Good ventilation is key*

It can't be stressed enough just how important adequate ventilation inside a chicken coop truly is. But it's vital, too, to appreciate that the type of ventilation is critical; drafts are bad, but a constant, gentle air circulation is good. As far as roosting chickens are concerned, drafts are bad news. They can cause unrest in the hen coop, induce stress among the birds and this can, in turn, lower their resistance to other problems or any underlying health conditions or parasitic burdens they may have. The ideal position for ventilation slots and holes is in the tops of the side walls and/or along the roof ridge. The idea is to set up gentle vortex currents that'll draw out the rising hot air from inside the coop, thus removing potentially harmful moisture and fumes. Keeping the coop clean obviously plays a key role in the effectiveness of good ventilation.

TIP 170: *Think about the long-term requirements*

One of the secrets of good poultry management—and it applies whatever size of flock you have—is to stay one step ahead in terms of the needs of your birds. It's always far better to recognize a potential problem before it strikes, and do something to prevent it, than being forced to react to it once it's happened. This general way of working with your birds applies to their housing just as much as it does to any other aspect. The last thing you want is to start having to deal with a downturn in egg production, behavioral issues, or ill health that's being prompted by problems with your chicken coop. It may be the wrong size, badly designed, drafty, leaking water, have poorly sited nest boxes, or be riddled with red mites because you've not been diligent with your antimite measures. Whatever the problem, though, you'll be playing catch-up as you search for the cause; this is never a good position.

RUN LOCATION

TIP 171: *Make life convenient, whatever the season*

Most people get started with chickens in the spring or summer, when the weather is fine and the temperatures high. Under those agreeable conditions "going that extra mile" isn't really an issue; we all like being outside when the sun is shining, after all. But trudging back and forth down slippery, muddy paths on a wet and cold February morning isn't quite so appealing. So, when planning the siting of your chicken enclosure, bear in mind the practicalities of getting to and from it when the weather's bad. Shorter routes are always going to be preferable to long ones when you're carrying full buckets to refill frozen waterers, or heavy sacks of feed. You might grow to curse the folly of tucking the hens away in that distant corner of your yard, just because someone thought it would be better to have them out of sight of those sitting on the terrace.

TIP 172: *Carefully assess the space available*

It's essential to be realistic about the space you have available, and to match it appropriately to the number and type of birds that you want to keep. Chickens thrive on space; they are foraging creatures that love nothing better than scratching around in search of insects and other tasty tidbits. However, the opposite is also true. Confine them in an area that's too small, and they'll start to suffer. The first thing to go will be the natural ground cover; the ground could be reduced to bare earth within a matter of days. Then, once the rains come, this will turn into a muddy mess that'll become increasingly dirty and present an ever-greater disease threat.

TIP 173: *Discuss the location of your chickens with any helpers*

If you need others to help with the routine chores associated with the birds—feeding, collecting eggs, cleaning—then it's important that everyone be clear about what's expected of them and that access to the run is made easy (and decide on the location as a whole group). It's vital that those involved fully appreciate that chickens are a commitment, that requires daily attention involving hands-on care to ensure their good health and welfare. There's hard work involved at times, too, and those who say they are going to do things must do them reliably and without fail. The birds in your care will be entirely dependent on you for their well-being. If you or others fail in your duties, they will start to suffer.

TIP 174: *And get your neighbors on board, too*

Keeping on good terms with your neighbors is important at the best of times, but it's doubly important if you're getting started with chickens. Unfortunately, local authorities are getting increasingly jumpy when it comes to complaints from residents about chicken-owning neighbors. It might be a crowing rooster, a smell, or the sighting of an odd rat that's triggered the complaint. Whatever the cause, though, local authorities often react in a high-handed manner, and it's typically the hen keeper who comes off worst. So bear this in mind as another factor to consider when getting started with poultry, especially if you're living in an urban area.

TIP 175: *Try a space-saving coop and run combination*

If you're concerned about the amount of damage that might be done to the rest of your yard if you let the hens free range, then you could consider buying or making a portable coop and run. This is a timber-and-wire framework that's used to contain the birds in a specific area, attached to the coop. This "chicken tractor" is designed to be moved to fresh ground on a daily basis. The idea is to clear the birds off any given piece of ground before any permanent damage is done to the vegetative cover and its root systems. This is a great system to operate, assuming you have enough space to move the run every day, and that you actually remember to do it.

TIP 176: *Steer clear of creating a messy eyesore*

In some respects, the smaller your yard, the harder you'll have to work with your poultry to keep everything looking spick and span. Where space isn't an issue, and you've been able to site the chicken enclosure somewhere relatively unobtrusive, then a bit of clutter here and there, and the odd patch of bare earth in the run, isn't much of an issue. However, when space is at a premium, and your birds are in full view, you're going to feel more inclined to keep things looking wonderful. This, of course, can work in the birds' favor in that, to maintain appearances, you'll be encouraged not to let your standards slip. Problems can arise, though, if you're tempted into taking on more birds than you really have space for. You'd be surprised just how quickly a gang of hens can reduce a sweet-smelling, grass-covered run to a barren, droppings-contaminated eyesore.

TIP 177: *Avoid especially damp areas*

Wet conditions are something that chickens find hard to deal with. We're not talking particularly about periods of rainy weather—although they don't like that, either—but wet conditions underfoot. If you allow a run to become a soggy, muddy mess, or you've mistakenly placed it in an area that's prone to flooding, then that's going to cause problems for your birds. Not only will they be generally miserable, but the increased risk to good health and welfare posed by being damp all the time is serious. There will be secondary effects for the chicken coop, too. The birds will be carrying in moisture all the time, which will cause the bedding material to deteriorate much faster than it otherwise would. It'll also promote a damp and humid atmosphere within the coop, which can be a major contributor to the onset of respiratory problems and other diseases.

TIP 178: *Protect special parts of your yard from the hens*

In practical terms, you must be prepared to write off the area of your yard that you turn into the chicken enclosure. Don't be under any illusion that your birds will go easy on your yard. Lots of keepers who, through limited space, are forced to keep their birds in a smaller enclosure than they would like, make up for this by letting the birds out to "free range" in the rest of the yard on a regular basis. This is a great thing to do—assuming the rest of the yard is secure and you don't have other pets that might cause problems—and the hens will simply love it. However, be prepared to fence off any vegetables you might have growing and any borders containing precious plants. Even the most docile hens can be pretty merciless with their scratching and pecking as they search for worms, slugs, beetles, and other insect life offering a tasty, high-protein snack.

TIP 179: *Chickens are happiest with some overhead cover*

Being descended from the Red Jungle Fowl—which, as you might have guessed, originated in the steamy jungles of Asia—chickens have an innate liking for overhead cover. It all stems from their natural fear of predators, especially airborne ones. This important aspect of their character is certainly something to bear in mind when planning your chicken enclosure. If you can incorporate trees, large bushes, and shrubs, then the birds will feel much more in their element. People who plunk a chicken coop down in the middle of a desolate field thinking they're giving the birds a healthy, fresh-air environment are actually doing them no favors at all. The big sky and lack of shelter will lead to the birds feeling insecure and probably spending more time than they should in the hen coop.

TIP 180: *Permanent runs need the right litter*

Of course, it's perfectly possible to keep two or three hens in a small (10 x 6½ft / 3 x 2m) run on a permanent basis, assuming that your husbandry and welfare routines are top notch. This sort of approach harks back to the wartime experience of many families, who, desperate for a supply of fresh eggs, would keep a handful of hens at home, often in a very limited space. Nowadays, our awareness of animal welfare issues is such that—quite rightly—there's much greater regard paid to general health and well-being. However, stressed hens won't lay well, so keeping them happy should be in everyone's interest. A permanent run is best fitted with a solid roof to give the birds protection from both the sun and the rain. This will also help preserve the bedding material. Ideally, this should be good-quality, dust-free softwood chippings. Bark chippings go moldy, while hardwood chippings can cause splinters.

SENSIBLE MEASURES

TIP 181: *Get water piped to your run*

However close to your coop the chicken run is located, a piped supply of water is a very desirable option. Doing this "by the book," with a buried supply connected to an upright pipe and faucet, is obviously going to be the most expensive option, and can suffer with freezing problems in the winter anyway. For those on a tighter budget, simply joining hose extensions together, and fitting a trigger-operated nozzle at the far end, makes for a perfectly adequate alternative supply method. Having a convenient source of running water close to the coop and run is a real advantage, and is certainly worth considering.

TIP 182: *Empty waterers to avoid freezing up*

Not only are frozen waterers a maddening nuisance, but they have potentially damaging consequences too. We all lead busy lives nowadays, balancing work with family life and leisure pursuits, so anything that saves a bit of time is always welcome. Discovering that your chickens' waterers are frozen solid is the last thing most of us need on busy winter mornings, as it always takes valuable minutes to clean out the ice and refill them. If you use plastic waterers, there's always the added risk that the expanding ice will have cracked the plastic, rendering the waterer useless. While covering the waterers at night with an old sack or bubble wrap might be enough to keep the ice at bay, the only way to be completely sure is to empty them the previous evening, once the birds have gone in to roost.

TIP 183: *Grid electricity makes a lot of sense*

We've already touched on the importance of keeping your electrified poultry netting in tip-top working order, so that its deterrent value remains strong at all times. If, like most newcomers to the hobby, you're running a battery-powered system, there's always a chink in your armor; batteries die, often without warning and always without alarm. For this reason it's important to remain vigilant with your testing, and also to have a second battery charged and ready for action at a moment's notice. Make sure the batteries you use are genuine marine/motorhome batteries (designed for use in motorhomes, golf carts, four-wheelers, etc.) from a good manufacturer. However, to overcome "battery-life anxiety" you could always equip your chicken run with mains power. Installation needn't be terribly expensive if you're prepared to do most of the work yourself, and the peace of mind and convenience that results can be worth the initial outlay.

TIP 184: *Take the strain out of carrying heavy loads*

A small wagon is one of the best poultry-related purchases. If you are fed up with regularly having to heft, for example, 44lb (20kg) bags of feed the 500ft (150m) from the trunk of your car to your storage near the coop, taking the plunge and buying a wagon will make life easier. The compressed bales of pine shavings should fit in it without a problem, and being able to pile it up with stuff and make just the one journey instead of multiple ones should make a genuine difference to the routine side of your chicken-keeping.

TIP 185: *Store the essentials close to the hens*

As well as having feed and shavings close at hand, it makes sense to keep other tools and equipment handily placed for use with your chickens, and a simple yard storage box is the obvious choice. There are plenty on the market these days, made from durable and weather-proof plastic. You can even get designs that double as a yard seat, which is even better. These can give you a great vantage point as you sip a mug of coffee and observe the scratchings of your contented hens. It's great to have somewhere dry to keep pellets, mixed corn, shavings, any nutritional supplements you may be using, plus your emergency first-aid kit. Having all this, together with a dustpan and hand broom, feed scoops, and an electric fence tester, makes life much simpler. However, you'll probably need to set a few mousetraps too, as it won't take long for the little pests to find their way inside.

TIP 186: *Always provide shade of some sort*

If your circumstances mean that it's not possible to provide your hens with any natural shade, then you'll need to think about what you can put together that'll do the same job. One obvious solution is the chicken coop itself. Raising this on extended legs to at least 10in (25cm) above ground level will give the birds enough space to get underneath, take shelter and probably create a dust bath, too. Getting the coop off the ground is a good idea anyway as it minimizes the risk of rat infestation, and all the best coops will be designed with this in mind nowadays. You might also consider building one or two field shelters, which simply consist of wooden legs topped with a plywood panel. The birds will certainly appreciate your efforts.

TIP 187: *Think about the orientation of your chicken coop*

Depending on the degree of exposure to the elements that your chicken enclosure experiences, it may be necessary to give some thought to the way in which the hen coop is positioned. For small-yard setups in sheltered urban situations, this won't be an issue, and the chicken coop can be positioned more or less however the keeper wishes. But in more exposed locations, where prevailing wind and rain may be a factor, it makes sense to orient the coop so that the pop-hole door is facing in the opposite direction to help keep the worst of the weather out. Larger hen coops and poultry sheds, which may feature a window, can benefit from being turned so that this allows in a decent amount of sunshine (this is especially beneficial for the birds during the winter and early spring).

TIP 188: *Keep a wary eye on the ground*

Regardless of how big your chicken enclosure may be, it's still perfectly possible for areas of it to become unsuitable for poultry. A prolonged period of rain, for example, can be all it needs to turn the birds' favorite area of the run into a muddy mess. You'll also have your habitual route to and from the hen coop, which will quickly become worn during spells of bad weather. All of this means that there's a constant need for you to keep an eye on the condition of the ground in your poultry pen and, if things get too bad, you need to be prepared and able to move things around to take the pressure off the most used areas.

TIP 189: *Grass is always the favorite ground cover*

There's no doubt that chickens love being out on sweet, fresh grass. A thick covering that's in good condition will contain a wealth of tasty treats for them to peck and scratch at. The action of doing so will keep them absorbed and contented, and the exercise will be wonderful for their all-around health and development too. For this reason, it's important to keep your runs as green as they possibly can be. Sometimes, though, the combination of the birds' activity and a run of bad weather contrive to make this impossible. This is what can lead to the need for a reseeding. Always pick a good-quality pasture mix for this. Your local supplier should be able to advise about what will grow best in your area.

TIP 190: *Keep the birds off damaged ground to give it time to recover*

Once ground in the run has become damaged, the only course of action is to give it time to repair itself. Much, of course, depends on how bad the damage to the ground cover actually is; if the roots have been scratched up and eaten, then re-seeding is the only way to recreate a quality ground cover again. However, if you just start to notice bald patches here and there, then the chances are that the grass will regenerate in these areas if given the opportunity to do so. This entails keeping the birds away from these areas until regrowth has occurred. With this in mind, it's handy to be able to subdivide your poultry enclosure, either with internal barricades made using something nonconductive, like wooden fencing, or by rerouting the main poultry netting so that the damaged area is fenced off.

TIP 191: *Hens can't have too much space*

Whether or not the space allocated to chickens in a run is adequate is always a bit of a debatable point. There are so many variables that it's almost impossible to offer a cut-and-dried verdict on what you should be aiming for. Factors such as the size, type, and number of birds involved, the quality of the land they're on, and the length of time they are to be kept on it are all relevant. Reading the many reference books already written about domestic poultry keeping will offer a range of suggestions but, as a very general rule of thumb, you could fairly safely work on the basis of allowing 27ft^2 (2.5m^2)of run area per bird.

TIP 192: *Be prepared to isolate birds*

There will be occasions when it becomes necessary to separate individual birds from the rest of your flock, and it's important that your setup allows for this flexibility. At its most basic, the use of another small hen coop (or even a rabbit hutch) may suffice, which can work well as a safe haven for a hen that's ill or needs a few days to recover from an injury. But, if you have the space, the best approach is to set up a completely separate run, with its own coop, feeder, and waterer. (See also tip 450.)

If ever you bring in new birds, these will need to be placed in isolation for a week or two before being introduced to the rest of your flock. These few weeks will give you a chance to assess the health of newcomers, and will prevent your existing birds from being needlessly exposed to anything nasty the new ones may be carrying. The same should apply to birds that have left your premises and been in contact with other birds—at a show, for example. Poultry exhibitions—where birds are caged side by side—can present a significant health challenge in terms of exposure to external parasites and disease.

TIP 193: *Make sure the fence is checked regularly*

Poultry run fences should not be left to their own devices. Inspections need to be made regularly and carefully to make sure that all's well. Spring and summer are the real "danger" times, when vegetation growth rates are at their highest, and battery-sapping short circuits—caused by the leaves and stalks of damp grass touching the current-carrying wires—are an almost constant threat. Spring storms, as well as fall and winter gales pose problems too, with the potential for falling branches and trees to wreak havoc. This is normally easier to spot but can still go unnoticed by those who don't take the trouble to carry out a perimeter check every day. You should also be checking at ground level for signs of digging and other disturbances that could signify unwanted attention of a predatory nature.

TIP 194: *Spare fence posts are always handy*

Even the most assiduous hen keeper will find it difficult to prevent electric poultry fences from starting to sag. Sooner or later, regardless of how regularly you adjust and retighten the ropes that support the corner posts, the netting just starts to droop. Whether it's an age thing, is related to its exposure to light, or whether the plastic simply stretches, the result is a droop that lowers the top line of the fence, and can bring the electrified wires into contact with surrounding greenery. By far the simplest solution to this problem is to add another of the specially designed fiberglass or polypropylene fence posts that are made for supporting this type of netting. The notches on the shaft work perfectly to drag the netting back up into line, and all's well with the world, once again. These posts are readily available online and should also be stocked by your local agricultural supplier. They're relatively inexpensive.

TIP 195: *Put your feeders and waterers on cinder blocks*

During the winter months, when the ground is naturally wetter, it makes a lot of sense to raise your feeders and waterers to help get them up and away from the mud that's bound to develop around their bases. You can buy specially made plastic waterer and feed stands, or use a cinder block or wood block to raise the height of the feeder and waterer.

TIP 196: *Don't feed local scavengers*

It can be a hard enough job to keep rodents under control at the best of times; the battle really is a constant one that vigilant poultry keepers must be aware of. One of the greatest attractions for rats and mice is poultry feed. There isn't much you can do about the hens' inevitably messy eating habits, although feeders with anti-scatter vanes will help reduce the amount of inadvertent scatter caused as the birds peck. With feed prices rising ever higher, you might want to think about moving the feeders out of the run at night (not into the chicken coop), so that you're not providing a free meal for any rodents or birds that fancy it. You may also try raising the height of the feeder, which reduces the amount that the hens from throw out, minimizing the amount of feed spillage. (See also tips 61–70.)

TIP 197: *Wild birds should dine at their own bird table*

The tendency for wild birds to feast at the poultry feeder, especially during the coldest winter months, has the potential to pose problems. There are diseases in the wild bird population that can affect chickens, and they're also carriers of parasites, both internal and external. Although the risks aren't usually that great, as a rule it's advisable to discourage this sort of freeloading. Many keepers leave their hens to feed on a free-choice basis, so feed is available to them whenever they want it. However, in cases where the local wild bird population latches on to this opportunity, it may be advisable to limit the hens' feeding time to just two or three supervised sessions a day. The wild birds will be reluctant to join in if you're on the prowl nearby.

TIP 198: *Scratch feed in the afternoons isn't such a bad idea*

Well-fed birds simply don't need to eat treats, and there could be health implications in feeding them too often. However, there is one exception: mixed corn can prove to be a very useful dietary supplement, scattered on the enclosure floor as a "scratch feed" during the late afternoon. Chickens love eating it, enjoy scratching for food, and the exercise involved in doing so helps keep them fit. Such a practice can be particularly worthwhile during the cold winter months, when the extra energy offered by the corn can really help the birds feel warm and content throughout the night. Take care not to give too much, though. About 1oz (40 ml) per bird should be quite sufficient.

TIP 199: *Spend time watching your hens' behavior*

Time simply spent standing beside your poultry enclosure and watching what the birds are doing is never wasted. It's amazing how quickly you can learn about how the different birds behave, and the way the group interacts as a whole. Developing this level of insight into their normal behavior is important because it'll help you to realize when things aren't right. One of the keys to good poultry husbandry is having the ability to nip problems in the bud. Anyway, apart from anything else, watching your birds is just a great way to spend twenty minutes.

TIP 200: *Tread carefully as you enter and leave*

One final common-sense measure relating to your poultry run involves nothing more than a few spare paving slabs, a couple of small pallets or two pieces of plywood sheeting. The idea is to give yourself "stepping stones" at the entry point to your poultry enclosure. If you've got the run surrounded by electrified netting, most keepers simply step over every time they pop in to collect eggs or open/close the pop-hole door. This is all well and good in the summer but, during a wet winter, your repeated footfalls will soon reduce the area to an unnecessarily muddy mess. Using slabs or wood to step on will prevent this and help keep your poultry pen just a bit more habitable for the birds.

THE RIGHT HOUSING

One of the most important choices you have to make when getting started with chickens is deciding which type of hen coop to buy. Not only will this probably represent the largest poultry-related investment you'll have to make, but it'll also have a profound effect on the health and welfare of the birds you keep. Get it right, and your birds will be assured a secure, spacious, and comfortable place in which to roost and lay their eggs. Mistakenly buy an unsuitable, substandard unit, though, and you could be sentencing the hens to a miserable, unhealthy, and even dangerous existence.

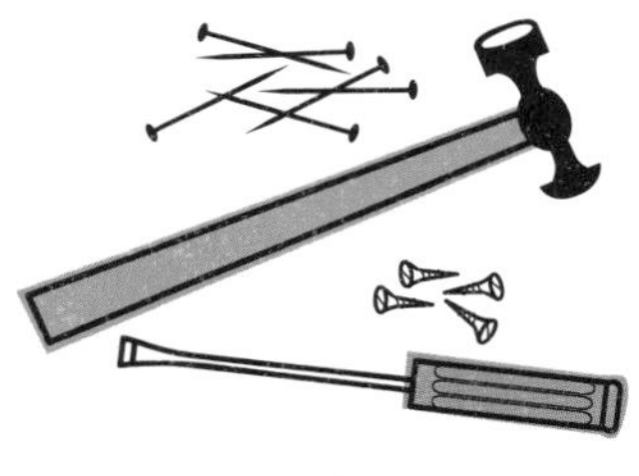

DESIGN CHOICES

TIP 201: *Pretty isn't always perfect*

Plenty of new chicken keepers buy their new hen coop largely on looks alone. People fall in love with a particular style and, once that decision has been made, they then look to work that style into the setup they're planning. Now, while this isn't necessarily the best basis on which to choose your chicken coop, it's what often happens in the real world, and the style that most choose is the traditional-looking, pitched-roof hen coop. With their playhouse looks and attractive detailing, these designs can represent something of a triumph of form over function. Nevertheless, people love them, and there are now countless producers making them. So if you get tempted into buying one of these pretty structures, please take the time to make sure that it has all the correct features, and that it'll perform the vital functions needed by its occupants.

TIP 202: *A utilitarian approach usually works*

At a more workmanlike level, the flat, or pent-roofed, poultry coop has always represented a truly functional option. The single-panel roof, which typically slopes to one side or the other ensuring water run-off, is cheaper to make but not as attractive to look at. But the birds inside care not, of course, as long as the coop offers them dry and draft-free security, plus enough space and adequate nest box facilities. Many pent-roofed coops also offer the benefit of a completely removable roof panel, which can greatly assist with cleaning and bird-catching operations.

TIP 203: *Watch the size if buying an ark-style coop*

Poultry arks were traditionally used out in the fields by semi-commercial operators. These simple structures that look like an A-frame coop were moved on an almost daily basis, so that the birds inside were never short of fresh grass to scratch in and eat. Nowadays you can buy ark-style units in various forms. Some will have a solid-walled coop section at one end, but the majority of the unit's length will be made up from wire-mesh-clad wood framework, ensuring that the birds have a bright and airy run area. Larger units can feature a double-decker style design, in which the coop is raised off the ground, thus maximizing the area over which the birds are able to pick.

TIP 204: *You could opt for the best of both worlds*

An increasingly popular housing option for those keepers with limited overall space is the coop/run combination. This, as the name suggests, offers a coop and attached run, usually in one, movable unit. These are available in a range of sizes suitable for varying numbers and types of birds, and offer a very convenient solution if your yard is small. As with all other types of housing, though, the quality of these units does vary from supplier to supplier. If you're planning to keep the unit on the same spot permanently, then this may not be a major issue but, if you want to move it on a regular basis, then its structure will need to be up to the job.

TIP 205: Sheds can make great chicken coops

There's no question that while a properly designed, purpose-built hen coop will do a great job for you and your birds, it's never going to be a cheap item to buy. So, if money is tight, then one option you might like to consider is converting a shed. You can work with an existing one (assuming it's in a convenient position), or buy a brand new one at a very reasonable price from your local hardware store. You'll have to put it together and adapt it, of course, but it should still work out to be considerably cheaper than an equivalently sized, off-the-shelf chicken coop. You'll need to add roosting, cut a pop-hole and make it closable, and build a removable, double or triple nest box unit, but these are just about all the alterations that'll be needed. A really big advantage of using a shed is that you'll be able to walk in and out of it with ease.

TIP 206: Plastic coops should last just about forever

There's still a good deal of resistance to plastic chicken coops, with the majority of new and existing hen keepers preferring the look of a traditional wooden unit. But attitudes are slowly changing. The reality, of course, is that plastic checks so many of the important boxes when it comes to poultry coop requirements. It doesn't rot or deteriorate with age. There's no maintenance requirement at all. It's easy to clean and can be more resistant to rodent gnawing. It's strong and, should a problem arise with a well-designed unit, it'll be simple to order and fit a new panel. One of the greatest advantages, though, is that the smooth-panel design of many of these plastic coops means that they offer very little refuge for red mites and other external parasites.

TIP 207: *There is the bricks and mortar option*

If you're lucky enough to live somewhere where you have access to an old barn or even an unused shed, these can provide ideal and extremely fox-proof housing for your hens. While you might not have the convenience of an enclosed run immediately outside the hen coop door, the typically solid construction should provide very secure and draft-free accommodation (perhaps after the odd bit of hole-plugging here and there!). Plus, as with using a converted shed, the volume of air within the structure should ensure that the environment inside remains fresh and healthy at all times, assuming you keep on top of your basic husbandry duties. Of course, you'll need to add suitable roosting and nest box facilities for the birds, with the latter probably requiring some degree of light control (loose-fitting burlap curtains are ideal) to keep them attractively dim inside for the laying hens.

TIP 208: *For a money-saving, satisfying option, do it yourself*

If you're handy with a saw and screwdriver, then the option of building your own chicken coop is something you should consider. There are books, magazines, and, of course, the internet, where you can easily find hen coop plans (and even instructional videos) that will enable you to build exactly the coop you're after. This ability to tailor the final result exactly to your needs is the really big advantage of the DIY option. So often, when buying a ready-made chicken coop, you're forced to settle for a degree of compromise due to the fixed options and limited choice. Designing and building your own coop will mean that you'll have the freedom to specify precisely what you want.

TIP 209: *Be mindful of inferior poultry coops*

The good news is that the everyday needs of a chicken, as far as housing is concerned, aren't that great. The bad news, however, is that many a modern coop design falls down in one significant respect or another. Poor design, cheap materials, and substandard construction can all seriously damage the ability of a chicken coop to meet the basic requirements. You would think that creating a safe, dry, and well-ventilated housing environment shouldn't be impossible but, regrettably, it quite often seems that it is. A significant part of the problem is fueled by economics, as producers strive to make their products ever cheaper to undercut their competitors. The problem for the inexperienced buyer is that, on the face of it, inferior hen coops can look every bit as good as superior ones. It's only with time—and exposure to a few months of bad winter weather—that the shortfalls or cost-cutting measures come to light. By this point, of course, it'll likely be too late, and the birds will be suffering.

TIP 210: *Always buy with care and attention*

It can't be stressed enough how important research and planning are when it comes to most aspects of getting started with chickens, including buying a hen coop. With this unit representing such a significant investment, it's very important that you get it right the first time. If you're a member of a local poultry club or society, then make use of the valuable experience you have on tap. Talk to people and get their views on housing types and setup options. Most people will be only too happy to offer their views, which will usually have been formed over many years spent keeping chickens. If you can, buy from a supplier that you can visit and where you can inspect the actual coops in their assembled forms. Take an experienced keeper with you for advice and moral support, if you feel you need it.

BUILDING MATERIALS AND STYLES

TIP 211: *Wood is great, up to a point*

Ever since people started keeping chickens at home as a convenient source of delicious, fresh eggs, wood has been the default choice as a coop-building material. Now, while it has a number of advantages—including being relatively cheap, easy to work with, and light—it has its downsides for this particular application, too. The fact that wood rots, warps, and splits with age has potentially serious consequences for people using it as a shelter-building material for their chickens. While pressure-treated lumber and wood-preserving treatments may prolong the life of the wood, you should be aware that the treatments administered to slow the rate of decay contain potentially hazardous chemicals.

TIP 212: *Take the environmentally friendly option*

Traditionally, plastic gets bad press, especially as far as the environment is concerned. However, an increasing number of innovative chicken-coop builders use composite lumber, which is made of plastic, wood fiber, and a binding agent, as their raw material of choice. With none of the inherent weaknesses of wood, plus a great recycling story to tell, many believe that this option will have an influential role in small livestock housing. On the downside, though, it is expensive, with composite chicken coops typically being on par with the very best wooden ones, in cost terms. But the material, while exceptionally easy to live with and maintenance free, struggles to become truly established against wood. Most people still seem to expect their chicken coops to look "natural," despite the associated problems.

TIP 213: *Avoid felt-covered roofs if you can*

Old-fashioned roofing felt has been used as a weather protector on the roofs of yard sheds and poultry coops for many years. It's just another of those traditional things that runs and runs, despite the problems it can cause from a poultry keeping point of view. Its continued use seems even more quaint given the fact that there are some excellent modern alternatives. But, there we are. It's a classic case of old habits die hard. The two big problems with traditional roofing felt are first that it hardens with age, then cracks and starts to leak and, second, it can provide an impossible-to-check refuge for that scourge of the hen coop, the tiny but oh-so-disruptive red mite. The fact that this felt is typically nailed down provides no barrier to the mite, which colonizes the space between it and the wooden panel on which it's mounted. This, of course, is impossible to check for without ripping the felt.

TIP 214: *Keep troublesome visitors at bay*

Hardware cloth—of various gauges and sizes—has played an important part in hen-coop construction for years. It's extremely useful for letting both air and light in, but keeping everything else out. Whether it's used over a door frame in summer, or over coop vents all year round, it's a sensible addition to help prevent wild birds and even adventurous rodents from getting into the coop when they're not wanted. Both have the potential to spread disease into your chicken flock. Rodents, of course, should never be attracted at any level as the secondary effects of their presence can be extremely costly.

TIP 215: Plywood is a really practical option

It's probably the expense of marine-grade plywood that prevents it from being used more widely on hen coops. Well, that and the fact that it's not a particularly attractive building material in its natural, unpainted form. But on a purely practical level, it's a stable, strong, and durable wood option than can be used to great effect. Typically, though, its use is restricted by most hen-coop manufacturers. Chicken coop floors, droppings boards, and, sometimes, roof panels can be made from marine ply, as can removable nest box units. But that's about as far as most manufacturers go with it, which is a shame. Its tough and long-lasting characteristics make it a potentially great coop-building material; what a shame it tends to be so expensive.

TIP 216: Tongue-and-groove to keep it simple

Some wooden chicken coops have side panels clad in boarding known as tongue-and-groove (T&G). This name refers to the simple method by which two flat pieces of wood can be joined, edge to edge. Each piece has a slot (the groove) cut all along one edge, and a thin ridge (the tongue) on the opposite edge. Slotting the tongue of one piece into the groove on another creates the joint, which is then typically tacked into place. It's a simple and very effective method of smoothly covering large areas or frameworks. So, as a general siding method, T&G works well.

TIP 217: *Fill in those cracks to fight the mite*

An old technique that was routinely carried out years ago, but seems to have declined now as common practice, was to paint the inside of the hen coop. The objective was twofold; to make it harder for pests like red mites to find refuge and to preserve/waterproof the wood to make cleaning easier and more effective. You can still do it today, of course, and there are a number of modern paint options that work very efficiently to seal up joints. The one thing you need to make sure of is that the product you use isn't toxic to the birds, but a specialist supplier should be able to advise on this.

TIP 218: *Hardware needs to last, too*

It's not much use specifying a coop that has been built using top-quality lumber if the manufacturer then skimps on the hardware, using cheap steel screws, hinges, and clasps. The constant exposure to the elements that a hen coop is forced to endure will soon expose any weakness in the overall design, and this will include the general deterioration of cheap hardware. Hinges will become stiff and screws will start to rust, weakening the overall structure. So always ensure that the coop you buy has been put together using stainless steel and/or galvanized screws and fixings if you want a unit that's going to stand the test of time.

TIP 219: Always ask about the prospects for coop repair

Wooden chicken coops are notoriously difficult to repair, which is why you always want to get yourself the best-quality unit you can afford in the first place. Damage caused by water (rotting wood), rodent damage (gnawed holes), or just good old-fashioned wear and tear will often occur in parts of the structure that aren't easy to fix or replace. So it's always worth asking the manufacturer/supplier whether they offer any sort of a repair service or replacement parts scheme. While the odd hole here and there can be patched quite easily, a split main door or a rotting corner in a floor panel can be a different matter. Wooden hen coops aren't generally designed to be taken apart, repaired, and then rebuilt to be just as good as before. Any reassurance that the supplier can give you on this point could be well worth having.

TIP 220: Safely dispose of old chicken coops

When you reach the end of the line with a wooden hen coop, or decide to upgrade to a larger or newer unit, think carefully about what you do with it. If you've battled with red mites during your time with the unit, then the chances are that the little creatures will still be in there. With this in mind, the safest option really is to burn it. Check your local regulations prior to lighting a match, however! Putting more birds inside will simply expose them to the mounting irritation, and passing it on to somebody else would do them no favors at all. If you want to leave the coop empty for a while, in the hope that the mites will die off through lack of food, be prepared for a long wait. These tenacious creatures can survive for at least eight months (some say ten) without food.

TROUBLESHOOTING COOP SIZE

TIP 221: *Don't cause an unnecessary egg hunt*

Keeping your hens contented, calm, and, at their healthiest and most productive, means providing them with the ideal environment. An integral part of that is their coop, which must be kept clean and inviting for them to use. Coops that are allowed to become damp and smelly will hardly represent an attractive proposition for a hen that's looking for somewhere inviting to lay her eggs. Failure in this respect can lead to you starting to find eggs laid in unusual places, which will almost certainly increase the likelihood of breakages and eggs that get lost altogether. It's important that your hens lay their eggs in the nest boxes provided, and that these are collected as soon after laying as possible. As you become familiar with your birds, you'll get to know when in the day they lay their eggs, and will be able to coincide your visits to the hen coop accordingly.

TIP 222: *Calculate chicken coop capacity accurately*

As your birds are likely to be spending at least 10 hours out of every 24 in their coop, it's essential that the environment in there is to their liking. One of the most basic requirements is space and yet this can be a slightly confusing issue. Coop suppliers can judge the capacity of their units by two different yardsticks; the amount of roosting space available or a per-bird allowance for floor space. Those working to the former assume that each bird will account for about 8in (20cm) of roost length, while the latter method requires $4ft^2$ ($60cm^2$) to be available per bird. Internal coop layout determines how these methods relate to each other, with the producers of small coops tending to prefer the roost length approach. When using the floor space calculation, however, there can be no debate.

TIP 223: *Deal with drafts straight away*

You might wonder at the distinction between ventilation and drafts, and whether it's important as far as hens are concerned. Well, it is. There's nothing more certain to make roosting hens unhappy at night than a drafty hen coop. Not only will it make them restless and stressed, but it can also trigger respiratory problems, which can lead to a serious downturn in health. It may also be sufficiently unsettling for the birds to prevent them from willingly entering the coop to roost at dusk. So it's important to deal with any draft issues there may be. If you're unsure about how to tackle things, get an experienced keeper to have a look for you.

TIP 224: *Make sure all the coop doors lock securely*

In all your efforts to get your poultry enclosure effectively fenced, don't forget the vital second line of defense, which is the security on your chicken coop. You need to be confident that, should a fox or other predator somehow get inside your main run, it won't be able to force its way into the hen coop itself. So it's important that the main door, the pop-hole, and the lid to the nest box (if an external unit is fitted) can all be fastened in such a way that they can't be "scratched" open. It's especially important to consider the intelligence and wonderfully dextrous little paws of raccoons! To this end, a lockable bolt can offer more security and peace of mind than a couple of simple metal thumb turns.

Chicken coop basics:

1. Sloping, removable roof panel; typically covered in roofing felt
2. Sturdy child-proof locks to secure main door
3. Covered vents; positioned high up for air exchange without drafts
4. Hinged nest box lid, also fitted with a metal lock
5. Strong locking bolt to secure pop-hole door
6. External nest box: maximizes coop space; makes egg collection easier
7. Typical tongue-and-groove wooden construction
8. Pop-hole; suitably sized for the breeds being kept
9. Entry and exit ramp, hinged to double as pop-hole door
10. Legs to raise coop off the ground; helps deter rodent activity
11. Main door: easy access for cleaning and catching birds
12. Sturdy main door hinges; the door fit must always remain good

TIP 225: *Inspect the things on the exterior that might otherwise go unnoticed*

Red mites are masters at keeping a low profile, and there are certainly a great number of inexperienced chicken keepers who don't realize they have a problem until it starts affecting their birds. With many wooden coop designs being so effective at providing countless nooks and crannies in which red mites can shelter and breed, it can be difficult to spot these tiny creatures during daylight hours, if you don't know what you're looking for. In fact, one of the easiest places to check is on the underside of the roosts. These should be removable, so take one out and inspect the bottom face, close to the ends. If you notice a gray, ash-like substance, or a mass of pinhead-sized, dark-colored dots, then you've found them and treatment should begin immediately.

TIP 226: *Tackle dampness without delay*

It's very important that anyone who keeps chickens appreciates the need to keep the inside of their hen coop dry. Insufficient ventilation and irregular cleaning out will create a humid and fetid atmosphere that's dangerous for the birds. But water leaks can pose problems, too. Much, as always, depends on the quality of the chicken coop you have and, to a lesser degree, its age. But the real problem is that water leaks take a gradual toll, so are easily missed in the early stages. Depending on how well the coop has been assembled, water can find its way in through joints between panels. If the roof is felted, this hardens and cracks with age, allowing a gradual trickle of water in that way. The problem is that this sort of regular soaking has one, inevitable consequence: rot. And if this is affecting the fabric of the coop, it can be eating away at the structure without anyone realizing. Keep a wary eye out for suspicious staining and damp patches, and don't just ignore them.

TIP 227: *Always match coop size to the birds being kept*

It may sound obvious, but lots of people make the mistake of forcing birds to live in chicken coops that simply aren't big enough for them. Most commonly it's a lack of headroom that's at the root of the problem. Somebody might start with bantams, then buy a few large fowl but never think about enlarging the coop too. Pop-holes that are too low (and sometimes too narrow) will create a needless obstacle for the birds and may cause damage to combs as well. A lack of headroom inside the coop, when the birds are roosting, is similarly undesirable, and may even prompt the birds to start roosting on the floor.

TIP 228: *A chicken coop should be raised on legs*

Not only will this make your regular coop-cleaning duties a little easier on your back, but raising the coop so that it's about 20in (50cm) above ground level will completely nullify the risk of rats forming a colony under the coop. This is a very real problem with coops that sit on, or only just above, the ground. Rats, in particular, are drawn to the warmth and security provided by the hen coop, not to mention the supply of food and water that's always nearby too. Some coop manufacturers have started adding legs on to their smaller units (or at least making leg kits available as optional extras). The key is to get the structure high enough off the ground to allow light underneath; this will be enough to deter rats from ever settling there.

TIP 229: *Large birds can't be expected to jump too high*

The height of the roosts within a coop can be an important issue with the heavier breeds, such as Wyandotte or Brahma. This is not so much to do with any inability the birds might have to get up on to the roost, but more to do with problems caused when they jump down again. The repeated shock of impact with the ground, when they haven't got much space to lessen it by flapping, can cause foot and leg problems. A painful condition known as bumblefoot can be a consequence, where an impact-related cut or puncture wound on the bottom of the foot becomes swollen, infected, and very painful. It's a slow problem to cure. Roosts for large, heavy birds need not be any higher than about 12in (30cm) above the floor.

TIP 230: *Think how it all looks from over the fence*

One final thing that it's sensible to consider if you're new to the hobby is how things are going to appear to your neighbors. It's all too easy, when you're brimming with the first flush of enthusiasm, to get very single-minded about your new passion, and forget the sensibilities of those on the outside, looking in. Everybody likes to keep on good terms with their neighbors; it's important to do so. But putting up fences and chicken coops left and right can be quite irritating if they're all backing on to your yard and you have no interest in poultry. So talk things through with those next door who may be affected by your "developments," and do all you can to address any concerns they may have. Get everyone around you on board, and you'll reap the rewards in the long run.

INTERIOR REQUIREMENTS

TIP 231: *Hens need a comfortable roost*

To roost comfortably at night, hens need to feel relaxed and secure. A lot of that has to do with the environment within the chicken coop, the number and type of birds in there, and the overall condition of the coop itself. However, at a more detailed level, the width and shape of roost the birds are using is a vital factor, too. It's important that the roost size makes it comfortable for their feet, and that means not having to grip on tightly all night. If the diameter is too small, then it'll be unpleasant for them to use, and might even break under their weight. The ideal is to have a roost that's wide enough to allow the birds to settle with their feet flat on it, so about 2in (4–5cm)—and a similar depth—should work well. Any sharp edges need to be rounded off as an additional comfort measure.

TIP 232: *Nest boxes need to be in the right place*

Hens are relatively fussy about where they lay their eggs and, as keepers, we need to pander to their needs in this respect if we want to maximize the production of undamaged eggs. Ideally, nest boxes should be positioned in the darkest part of the hen coop. The number of boxes is important too, with an allowance of one box for every three hens. They should be lined with the same material that's used on the main floor in the coop (clean, softwood shavings are generally regarded as the best option). Externally mounted nest boxes, with a lid that can be opened from outside the chicken coop, are the most convenient option from the keeper's point of view.

TIP 233: *Discourage your hens from roosting in the nest boxes*

It's a nuisance to have your hens sleeping in the nest boxes for a number of reasons, so the practice (which quickly becomes a habit) needs to be discouraged. Hens will soil the nest box during the night and, if the droppings aren't then cleared up on a daily basis, they'll lead to an increase in dirty eggshells. This results in more work for you. Common causes for nest-box roosting are the boxes themselves being badly positioned within the coop, overcrowding, and insufficient or inadequate roosting. It's important that nest boxes are set at a lower height than the roosts, as hens will instinctively choose the highest available roosting point. So, if the nest boxes are positioned high up on the side wall of a chicken coop, that's where the birds are likely to go for the night.

TIP 234: *Make sure you can secure the pop-hole door*

Pop-hole doors open in a variety of ways: they can hinge to the side, drop down to form the entry ramp, or slide to one side or up. The important thing with all of them is that they are secure when closed to keep predators at bay. They need to be a good fit, too. Mice and rats can gain entry through surprisingly small gaps, so if the pop-hole door isn't a particularly good fit, these unwanted night visitors will get in. Mechanisms that allow the door to be pinned, bolted, or clipped shut make a lot of sense, as do sliding doors that close into a groove that prevents them from being hooked open by a well-placed claw.

TIP 235: *Check for potential dangers inside the hen coop*

It's always well worth spending a few minutes carefully inspecting the interior of your new chicken coop, checking for anything that might cause the birds injury. Splinters of wood and protruding screw tips are the sort of little oversights that have the potential to cause trouble. These defects shouldn't occur in a top-quality hen coop, but may well be present in something that's been put together by a less-experienced producer. Wine bottle corks are very useful for winding on to protruding screw tips as a quick fix, and dealing with splinters is nothing that a block and sandpaper won't be able to sort out in a matter of moments.

TIP 236: *Interior lighting needs to be safe*

Some keepers install electric lighting in their chicken coops, either for their own convenience during the long, dark days of winter, or as a controlled way of prolonging daylight hours and thus maximizing laying performance. However, whatever your reason for introducing electricity into your hen coop, it's essential that it be safe. For obvious reasons, any wiring needs to be kept well out of the birds' pecking range, and the possibility that the insulation could be attacked by rodents must be considered too. Exposed filament bulbs generate a lot of heat and, in an environment that may have become dusty and cobwebby, this can pose a fire risk. For this reason, cool-running, energy-saving bulbs make a lot of sense, even though their performance isn't quite the same.

TIP 237: *If there's an easy-clean option, go for it*

Cleaning out a chicken coop is one of those jobs that very quickly loses any novelty that it might have had at the start. So the easier and quicker a hen coop is to clean, the better you'll like it. These benefits will be particularly appreciated during the freezing cold and wet days of January and February. A large door for ease of access, the presence of a big droppings board, and the absence of an irritating sill preventing the floor litter being conveniently swept out are all desirable features to look for.

TIP 238: *Chicken coop floors don't have to be solid*

The fact that virtually every domestic-type hen coop you'll see these days has a solid floor is slightly misleading, because it's not the only option. In days gone by, plenty of larger chicken coops would feature a floor made of wooden slats. Not only did this negate the need for bedding and all the work and expense associated with its use, but it also allowed the birds' droppings to fall straight through and on to the ground below. So, by moving the coop on a regular basis, it acted as a very ecofriendly fertilizing machine! Hardware cloth is another durable flooring option, which has the added benefit of being completely rodent proof. Of course, on the downside, an "open" floor like this isn't as pampering for the birds but, then again, in some respects, this may be regarded as no bad thing.

TIP 239: *Minimize your egg losses*

Not all hens are light on their feet and agile; some are just plain clumsy, which is one of the reasons you can find broken or cracked eggs in your nest box. It's bad to let this continue, as it encourages egg eating among the birds, and this destructive habit is a hard one to break. One solution to this sort of trouble is to use a roll-away nest box. These simple devices rely on nothing more technical than a gently sloping floor, which, once an egg has been laid, causes it to roll forwards or backwards into a covered compartment where it's safe from even the most determined pecking.

TIP 240: *Removable components are easier to clean*

Nowadays, thanks to the revolution in manufacturing that modern plastics have brought with them, it's possible to buy handy roost sets and nest-box units made out of the stuff. These can be placed wherever you want inside your hen coop, and are especially useful for those converting a shed for use as a chicken coop. Being simplicity itself to remove, and made from nonabsorbent hard plastic, they are a cinch to clean and disinfect. Plus they are far less welcoming to undesirables such as red mites. You can even buy externally mountable nest boxes, which, with the addition of an appropriately sized hole in the side of your shed conversion, can be mounted securely in place to aid convenient egg collection.

MAINTENANCE

TIP 241: *Keep a careful eye on chicken coop condition*

Too many keepers treat their hen coop as a once-and-done component; they buy it, set it down in the chicken enclosure, and then tend to ignore it. This really is very unwise, especially if you've got a wooden coop that, in theory, can start deteriorating from the moment you put it outside. So the name of the game is vigilance at all times. Don't ignore your hen coop; it probably represents your most significant poultry-related investment. You need to be able to act quickly and nip potential problems in the bud.

TIP 242: *Protection starts at the top, with the roof*

The effectiveness of just about any chicken coop will be governed by the condition of its roof. Weaknesses in the structure or weather-proofing will be exploited by the weather, to the detriment of the birds inside. Rather ironically, it's the chickens themselves that can make matters worse (together with wild birds). Roof panels that are covered in felt become vulnerable as they age; the covering becomes increasingly brittle and eventually cracks. However, birds can speed this deterioration thanks to their habit of pecking and scratching at the covering. So keep an eye out for wrinkles that develop in the surface and any nicks that start to get enlarged by the birds. Wooden roofs tend to be more durable, although be on your guard for splits in the wood or warping, which might start to allow water inside.

TIP 243: *Re-treat your hen coop once a year*

Once your wooden poultry coop gets to two or three years old, you'll have to start treating it to a coating of wood preservative (inside and out) to maintain its resistance to the elements. Be aware that some wood preservatives may be toxic and the fumes may be hazardous to birds' delicate respiratory systems. To do the job properly you'll need to re-house your hens for a few days in temporary accommodation, to give whichever product you use the chance to dry completely before re-introducing the birds. There are plenty of good wood preservative treatments on the market these days, and you'll need to take the time to coat both the inside and the outside thoroughly. It also makes sense to tackle this job in the late spring or summer, when the weather is both warm and dry.

TIP 244: *Make poultry-related repairs as soon as possible*

It's important to appreciate that chickens are a commitment, and that they are just as deserving of our care and attention as other things that may, on the face of it, seem more pressing. But lifestyle pressures can mean that things get put off, especially in the poultry pen. Bad weather can be another major deterrent to getting there and doing stuff that needs doing. But it can be a risky policy. Putting off trimming the grass around the base of the electrified netting, delaying cutting back that overhanging shrub, or failing to tighten the hinge on the pop-hole door can all have serious consequences for your birds.

TIP 245: *Established cleaning routines save time*

If you can get the rest of the family involved with the husbandry side of looking after the chickens, then all's well and good. It won't suit all, of course, and there's no point in forcing youngsters to do chicken-related chores that they're not interested in, because you need to trust that they'll be doing it right, every time. So, if there's any doubt, the best thing is to tackle the work yourself, safe in the knowledge that it'll actually get done. Daily "poo picking" from the coop makes a lot of sense and, if your birds have a small run, then removing poop from the ground in there can help too. In their natural, raw state, chicken droppings can be quite acidic, so their rapid buildup on specific areas of ground isn't ideal and can actually be detrimental to plant growth.

TIP 246: *Do a major clean-up every six months*

While the weekly routine of sweeping out the hen coop, replacing the bedding, and treating with whatever disinfectant powder you choose to use is good practice, there's still a need to give the coop a thorough cleaning, perhaps twice a year. By "thorough," use a jet wash to blast detergent into all the cracks and crevices to get the whole structure as clean as you possibly can. Not only will this involve dismantling as much as you reasonably can, but you'll also have to ensure that the coop dries out completely before the birds are allowed back in. Doing the job properly will require the provision of some temporary accommodation that your birds can use while their coop is drying.

TIP 247: *Sometimes it's just easier to replace things*

The trouble with a lot of poultry equipment is that repairs are very tricky. On a wooden chicken coop, if it's not an easy-to-get-at part, such as a main door, nest box lid, or bird ramp, then it's unlikely that the supplier will be able to do much about it. Those who are also manufacturers may be more able and inclined to help, but don't count on it. It's lucky that poultry keeping isn't a particularly "equipment-heavy" kind of hobby. The basic requirements, apart from the coop, fencing, feeder, and waterer, are all pretty ordinary, multipurpose things. So splurging on new replacements for most of them every now and then shouldn't normally break the bank.

TIP 248: *Don't develop a poultry drinking problem*

Waterers should be thoroughly cleaned out every few weeks to prevent green algae buildup. Use a citricidal disinfectant and a scrubbing brush if you want a completely natural approach; alternatively, you can use solutions of iodine, chlorine, and phenol.

TIP 249: *Use a modern solution to an old problem*

Traditional, felt-covered roof panels on hen coops can cause problems; increasing numbers of British chicken keepers are switching to a relatively new roofing solution, using a product called Onduline. This is an extremely tough, lightweight, corrugated roofing material that's manufactured using a base board produced from recycled cellulose fibers that is saturated with bitumen under intense pressure and heat. The benefits include the fact that it's easy to cut, shape, and secure, has a 15-year water-proofing guarantee, comes in a range of colors (including black and green), doesn't rot or become brittle, and is flexible. Onduline recently acquired Tallant, a major supplier of lightweight roofing in the United States, to enter the US market.

TIP 250: *Keep netting well connected*

By now you should all appreciate the importance of keeping an electrified poultry netting enclosure fence in tip-top condition; a system that's operating below par simply won't provide a practical deterrent to those disreputable characters on the outside. One quite common cause of a reduction in performance is broken wires. These are very thin—they need to be, so that they can be woven into the netting fabric—and consequently they're quite easily damaged. In fact, careless trimming, or getting too close with a brush cutter as you battle with the vegetation, will slice through the whole thing. Inadvertent wire chopping needs to be dealt with. Fortunately, most good poultry netting setups are supplied with a repair kit, so use it.

FEEDING AND WELFARE

Although chickens are actually remarkably resilient creatures—much more so than you might imagine given their delicate outward appearance—they remain completely dependent upon their keeper for good health. Their ability to disguise the onset of disease can work against them in the domestic environment, sometimes making it difficult for novice keepers to spot when things are starting to go wrong. All this makes it doubly important that, as a responsible owner, you do all you can to feed and care for your birds as well as you possibly can.

HEALTH BASICS

TIP 251: *Keep them healthy, keep them laying*

It's in your own best interests to keep your hens in the best possible condition because, unless their health is tip-top, they're going to underperform in all respects. If you've hatched birds and are raising them from chicks, challenges to their health—even underlying, minor ones—will have a detrimental effect on overall growth rates. Young chickens grow best during the spring and summer, with the sun on their backs. If you miss this important season because growth rates are slower than they should be, then these birds will often never catch up and will finish up as poor specimens. Laying hens need to be in full health to give of their best. Hybrid layers and the best utility strains of the pure breeds will only be physically capable of rewarding you with the expected production if there's nothing going on in the background that's sapping energy.

TIP 252: *Chickens are tougher than you might imagine*

Chickens are remarkably resilient creatures and also great at disguising health problems. When problems start to become apparent due to a hen's behavior—it's lethargic, hunched, stops laying, and stands around with its feathers slightly fluffed up—you know that, actually, things are quite bad for the bird. In most cases, assistance from a vet will be required. It's possible that the bird will get better on its own, but it may simply get worse. You must consider its overall welfare at all times: if you believe it's suffering, then take action on the advice of a professional.

TIP 253: *Try not to let your hens get stressed*

The susceptibility of chickens to stress certainly represents a weak link in their generally impressive constitutional armor. Although some pure breeds can be a little more highly strung (rather like pedigree dogs), the average hybrid layer will sail through life carefree. However, all chickens can be dragged down by the effects of stress, which can be caused by things that we might otherwise regard as trivial. Next door's barking dog, for example, can be enough to trigger a problem, or the overenthusiastic attentions of children who think it's fun to regularly chase hens around their pen when nobody's watching. The unfortunate consequence of stress in chickens is that it has the effect of lowering their resistance to external disease, as well as making them more susceptible to any underlying health issues there may be.

TIP 254: *Remember, chickens get sniffles too*

Like we humans, chickens can suffer from a range of respiratory diseases, the main signs of which include breathing difficulties such as gasping and sneezing, plus wheezing noises, gurgles, and rattles from within. In fact, respiratory problems are among the most common ailments affecting domestic chickens. Thankfully, problems of this sort are rarely fatal, although, because of this perceived lack of severity, birds often go untreated. Respiratory diseases can be caused by virus (infectious bronchitis), bacteria (infectious coryza), mycoplasma (sinusitis), molds and fungi (aspergillosis), nutritional deficiency, parasites (gapeworm), and tumors. "Rattling" is a symptom of respiratory troubles; the sound may come from the upper end of the respiratory tract, or be more deep-seated in the air sacs. In some cases birds will recover without help but in most it is fairly sensible to seek veterinary assistance. (See also tips 421–430.)

TIP 255: *Don't let the grass grow too long*

"Impacted crop" is a condition where the crop—a pouch-like food storage point at the base of the neck—becomes swollen and tightly impacted with material such as long, tough grass; feathers; wood shavings; and straw. This is potentially serious as the bird stops eating because the passage of food is impeded. Weight loss follows together with a generally depressed state. White, liquid droppings will be seen from the sufferer, and the diagnosis can be confirmed by feeling the hardness of the crop. A home remedy is to dose the bird with a teaspoon of olive oil, then gently massaging the crop in an attempt to break down the contents. If this approach fails to clear the blockage, then the stubborn contents of the crop will have to be removed surgically. So keep run grass short and don't feed long vegetable peelings.

TIP 256: *Viral infections can be a serious issue for hens*

Viral diseases have long provided a major cause for concern among poultry keepers, and there are seven main ones to be aware of: Newcastle disease, Marek's disease, infectious laryngotracheitis, infectious bronchitis, egg drop syndrome, avian encephalomyelitis, and fowl pox. These all sound pretty nasty and, indeed, they are! The situation isn't helped by the fact that treatment for them all is relatively unsatisfactory. All can be vaccinated against and you'll find that most hybrids will have been protected in this way at an early age. Pure breeds, on the other hand, are unlikely to have been given any such protection. Vaccines are generally very expensive, and administering them effectively can be a specialized business. Hence this tends to be the domain of the commercial producers. Domestic keepers rarely treat their young birds in this way.

TIP 257: Droppings can tell you a great deal

A healthy hen produces a distinctive kind of dropping, which you'll probably most easily find on the coop floor. It'll be fairly firm, rounded, and with distinct white- and dark-colored sections. The larger of the two portions—the dark one—can be black, brown, or gray and is the solid waste, while the white section is the urine. Droppings act as a great indicator for when all is not well with a hen and are a very useful pointer towards what's ailing the bird. Variations also occur quite naturally. Intestinal lining gets shed every now and then, leading to a pink or red coloring. Coral-colored droppings are often produced overnight, and are quite normal too. Mustard-colored, foul-smelling droppings are caecal droppings, and are typically expelled every ninth or tenth dropping. However, watery diarrhea can be a sign of enteritis or anemia; bloody diarrhea points towards coccidiosis, while yellow-colored can suggest respiratory disease or worms. Fecal analysis is the only way to be sure.

TIP 258: Parasite problems are common

The parasites that affect chickens either ply their trade on the outside of the bird (ectoparasites), or go about their business on the inside (endoparasites). Typical of the creatures you might find on a bird's body are lice, ticks, and mites; infestations of any of these will put a drain on resources; and cause irritation, a reduction in laying performance; and, in really severe cases, can prove fatal. The most commonly encountered internal parasites are *Ascaridia galli*, the large intestinal roundworm; *Heterakis gallinae*, the caecal worm; *Capillaria* or hairworms, found in various parts of the digestive tract; *Syngamus trachea* or gapeworm, found in the wind-pipe; and *Davainea proglottina* or chicken tapeworm. Action should always be taken against parasites. There are plenty of treatments available for both external and internals types (see also chapter 9), but do consult your vet.

TIP 259: *Leg problems can be really irritating*

Chickens can suffer with an extremely irritating condition called scaly leg, which is caused by a tiny parasite that burrows under the scales on a bird's leg, where it then lives and breeds. The condition tends to be more common in older birds, and those with feathered legs, and can be recognized by the appearance of lifted scales and dry, gray, and crusty deposits. As a result, the shanks can become enlarged and hot to the touch. This parasite causes intense irritation for the bird and, if left untreated, the feet will become deformed so much that walking will become difficult. This condition can be treated by slathering the legs with petroleum jelly and cleaning bedding and roosts. Eradication is likely to be a long process requiring patience and perseverance. Never pull off the lifted scales in the hope of improving things; this will be extremely painful for the bird. Instead, they must be left to fall off and regrow at the next annual molt.

TIP 260: *Be aware of "bird flu," and how to best protect from it*

Avian influenza is a normally fatal disease for chickens. It's spread by all birds, making it extremely difficult to contain; migratory waterfowl are particularly effective carriers. The signs of infection are essentially limited to a "blueing" of the bird's comb and wattles, followed by sudden death. To prevent spreading, the infected birds must be kept isolated from all other birds, using enclosed runs with wild bird-proof netting and (ideally) solid roofs. Successfully treated birds will be carriers.

THE PECKING ORDER

TIP 261: *There will always be a "top dog" in your flock*

It's important not to dwell too much on the negative aspects associated with keeping chickens but, at the same time, we all need to be aware that bad things can and do happen every now and then. After all, you're bringing together a group of birds (often of differing breeds) and expecting them to live harmoniously and without incident in a false environment created by you. An important part of the "living together" process, as far as chickens are concerned, is the "pecking order," which, in practical terms, is a ranking of the birds in order of dominance. It's a primitive system based essentially on power and strength of character; those at the bottom can have a tough time, especially if space is tighter than it should be.

TIP 262: *Prevent hen disagreements as much as possible*

Typically, disagreements between chickens are short-lived affairs. The dominant bird will exert its authority with a flap of the wings, a petulant cluck, and maybe even a peck and this will usually be enough to send the subservient bird scurrying off out of range. On occasions, though, the aggression doesn't stop there and bully/victim situations can develop. This is something that keepers need to be ever watchful for because, if allowed to continue unchecked, these situations can turn nasty. Any hen that's being picked on by one of more of its flock mates will feel stressed. It may be prevented from feeding properly, or even from getting into the coop when it wants to lay or roost. This is why it's so important to observe your birds' behavior on a regular basis. Ultimately, a bird that's being endlessly bullied may need to be isolated from the others, for its own protection.

TIP 263: *Bored birds may peck at each other*

Feather-pecking is another of the poultry "vices" that needs to be watched for and dealt with sooner rather than later. It can be triggered by all manner of factors, but bored birds and those confined in a space that's too small for them can be particularly susceptible. It's a very destructive habit that can also be prompted during the molt (when new feathers are just starting to come in) and, once birds get used to doing it, it can be very hard to stop them. Problems arise if the bird having its feathers pecked sustains an injury that draws blood, as this then becomes an irresistible target for others in the flock.

TIP 264: *Never forget that hens love eating eggs too*

Egg eating can be one of the most irritating of the poultry vices, simply because it represents such a waste of good eggs. Given the opportunity, hens love nothing better than eating the contents of a fresh egg. Normally, of course, they aren't given the chance to do so but, if an egg gets accidentally broken in the nest box by a clumsy hen, or a soft-shelled one is laid, then they'll waste little time in taking advantage. The trouble is that it doesn't usually end there. They quickly learn to break good eggs to get at the contents (those laid on the coop floor are always most vulnerable), and the problem simply escalates. It's a tough habit to break! In practice, emphasis has to be placed on prevention. Ensure there are adequate nest boxes, that they are at least 18in (45cm) above floor level, and as dark as possible. Persistent offenders will need to be removed. (See also tip 444.)

TIP 265: *Be prepared for some nastiness now and then*

There's no doubt that chickens can be vicious creatures; the carnivorous side to their character ensures that, if push comes to shove and the opponent in the firing line is much smaller than they are, well, there's only going to be one winner. Regrettably, though, this ferociousness can be turned on their own kind, and often for no apparent reason. Literally, it can strike out of the blue and with devastating effect. The first thing you'll know about it is that you'll find the bloody remains of a bird on the floor in the morning. It can be triggered as a result of an escalation of bullying behavior, or caused by the "blood lust" that can be prompted when one of the flock receives a flesh wound that bleeds. Once again, intervention is required by the vigilant keeper to nip these sorts of problems in the bud.

TIP 266: *Mycoplasma can be triggered by poor husbandry*

This is a serious disease that affects a bird's respiratory system, and there are a number of forms that can strike chickens. Many are present naturally in the birds, only becoming apparent when something else, like a virus or stress, weakens the bird. There are other types that are pathogenic, and affect otherwise healthy birds. Even these, though, normally require some other predisposing factor, such as poor air quality, poor hygiene, or high levels of stress, before taking hold. The type that causes the most problems for chickens (*Mycoplasma gallisepticum*) is spread by egg transmission, aerosol infection (from infected droplets of moisture from sneezing or coughing birds), and by direct contact with infected and/or carrier birds. It causes chronic respiratory disease and air sac syndrome. Birds affected will sneeze, cough, and produce some nasal discharge. Antibiotics can be used to fight the condition, although resistance is increasing. (See also tips 421–430.)

TIP 267: *Take care when rearing young chickens*

Coccidiosis is a potentially fatal disease that mainly affects young chickens (between three and six weeks old), but can strike adults too. Typical signs include depression, huddling, feather fluffing, white soiling of the vent, diarrhea, and paleness. All hens will possess some "coccidia"—a low level of infection is common. It's caused by a protozoan (a single-cell parasite), and is spread among the flock via feces. It can survive out of a bird (inside a hen coop) for more than a year. Factors that increase the risk of "cocci" include overcrowding, keeping young chicks where older birds have been, and rearing birds in a warm, wet, and underventilated environment. Chicks have no immunity from the hen, and only gradually acquire resistance after low-level exposure, going on to become fully immune to that one strain at about seven weeks old. Chick starter can be bought containing coccidiostat, but good husbandry is the real key.

TIP 268: *Be on your guard against worms in your birds*

There are many different types of worm that can infect chickens. Lots don't cause problems, but some do. For example, *Capillaria contorta* and *Capillaria annulata*, are both hair-like worms found in the esophagus and crop. The larvae develop inside earthworms before becoming infectious to chickens. They bury their heads into the lining of the mouth, crop, and/or esophagus, causing inflammation and a thick layer of dead and dying cells that reduces the normal gut contractions. The results are a loss of appetite and weight. Tetrameres affect free-range chickens, with their larvae developing in grasshoppers, beetles, and crustaceans. Once eaten, they infect the bird's stomach. The female worms are bloodsuckers, so can cause anemia. Treatment for all worms is with the use of a good wormer product. Consult your vet for up-to-date advice about what's best to use. (See also tips 411–420.)

TIP 269: *Avoid Marek's disease by removing "dander"*

While controlled commercially with vaccine, Marek's disease still causes problems—and fatalities—among domestic flocks. The Sebright and Silkie are the two most susceptible pure breeds. The disease is caused by a herpes virus and is spread by feather dander—microscopic particles that can travel miles on the wind. Carrier birds shed Marek's virus with no signs of infection themselves. The disease causes the growth of tumors in the nervous system, resulting in various degrees of paralysis—legs, wings, and head are typically affected. Birds that don't die become carriers, and may succumb in the future if they become stressed. The best approach is to breed for resistance, but the level of viral challenge can also be reduced by sweeping or vacuuming (wetting first) the poultry coop daily, to remove as much of the dander as possible.

TIP 270: *Always handle your laying hens with great care*

If you're rough when handling your hens, then there's a chance that you could break an egg inside a bird, which can have disastrous consequences. Egg peritonitis is an inflammation of the peritoneum (the thin membrane surrounding and supporting abdominal organs, and lining the abdominal cavity), caused by bacterial infection. The most likely cause is egg material entering the abdomen instead of the oviduct, or through impaction or rupture of this organ. In acute cases, a bird will likely die before any symptoms are noted. A postmortem examination will likely reveal putrid yolk material in the abdomen (and a horrible smell). This condition is more likely to occur in birds that are just coming into lay, and those in an overweight condition. Regrettably, there is no treatment.

   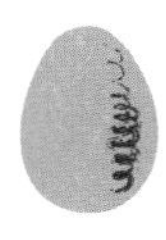 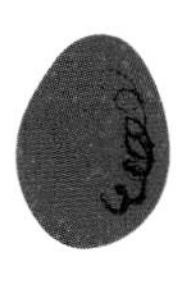 

KEEPING HENS AMUSED

TIP 271: *Give hens enough space to have fun*

We've already established that stressed hens can soon become unhealthy hens, as anxiety actually weakens their immune system and increases the likelihood of them succumbing to an external disease, or some pre-existing but underlying medical condition. It's essential to keep your hens actively involved in everyday life and, at the core of that, is giving them enough space to flap, scratch, and peck naturally to their hearts' content. In cases where space is limited then you must be doubly sure that you don't fall short in two vital respects. For a start, don't overstock with too many birds. There really is no excuse for this. Second, if your run is small, then make sure that those inside have plenty to keep them occupied at all times.

TIP 272: *You've got to stimulate to accumulate*

If you reward your hens with the sort of environment in which they feel happy and safe yet, at the same time, are stimulated and active, then they will repay you with good health, trouble-free keeping, and a fantastic supply of tasty, fresh eggs. Good poultry keeping is all about quality of life; both the birds and yours. The last thing you want—especially if you're new to the hobby—is to be continually fire fighting as you tackle problem after problem with the birds. It's not good for you, and certainly isn't beneficial for the birds. Their needs aren't great, but if you cut corners on the basics, then you're inviting trouble.

TIP 273: *Encourage natural behavior among your hens*

A chicken is never happier than when it's out, with the warm sun on its back, scratching around in a tireless search for tasty things to eat. You can encourage this by providing a run environment that includes a variety of ground cover and natural features. You can also pique their interest at certain times of the day by scattering layer pellets or mixed corn in different areas of their enclosure. Don't overdo things though. It's important that everything gets eaten so that food isn't hanging around to attract wild birds and rodents. But at the same time, you don't want to feed so much extra that the birds start to get fat. Overweight hens don't lay well and suffer with other health issues, as you might imagine.

TIP 274: *Your hens will go crazy for fresh greens*

Although hens love eating ants, worms, beetles, and any other creepy crawlies they can lay their beaks on, they're also big fans of fresh greens. If you can combine this passion with something a little different, then so much the better. One way to do this is to suspend fresh greens on a string in their run, setting them at a level that's just above head height. The birds will have to work a little for their pleasure, by jumping up as they feed, which will be very beneficial for their muscle tone as well as all-round health.

TIP 275: *Though some fresh greens are better for compost*

Birds will devour most tender-leaved plants when fresh, even nettles. As a rule, they're usually pretty aware about what they can and can't eat. However, there's no point in taking needless risks and there are certainly a few plants to avoid tempting them with. Things that could do chickens harm if eaten include: delphinium, ivy, foxglove, hyacinth, hydrangea, oleander, rhododendron, and rhubarb.

TIP 276: *Make friends with your birds*

It's very important that your birds feel comfortable around you; they need to be relaxed in your presence. Chickens are very habitual creatures. They love routine and will quickly learn when and where they are normally fed. They also rapidly become familiar and relaxed with their keepers, assuming you're not always spooking them, holding them badly, and making loud noises and sudden movements. If you take the time to sit around in the pen and allow the birds to develop confidence in you, then you'll find all aspects of the daily husbandry routine much easier and more satisfying. It's all common sense really, but the sooner you can get your birds to feel an affinity with you, the better.

TIP 277: *Hens are sociable, so give them company*

Like most birds, chickens are flock animals that enjoy the interactions associated with being in a group. They feel happiest—and most secure—among their own, which is why it's important never to keep them singly. One hen on her own will be a miserable creature from the beginning, so birds need to be bought in twos or threes. Don't buy one first, then add another couple a few weeks later because this could stir up trouble when the new birds meet the old one. It's always best to buy your birds as one same-age group to eliminate the prospect of "introductory" trouble.

TIP 278: *Let them out whenever you can*

If your situation means that you're only able to keep a few birds in a fixed run because you have a small yard, the birds will still benefit from being let out every now and then. Anything you can do to spark their interest will help, even if your yard is for the most part paved with just a few tubs or raised beds, the hens will still relish the chance to pick around in what soil there is around whatever plants there are. Incidentally, if you're worried about the birds escaping when they're let out to roam the garden, just snip about 2in (5cm) off the tips of the flight feathers on one wing, to unbalance the bird and prevent it from flying. This will mean that you'll need to be on hand when they're out, just in case a potential aggressor comes calling.

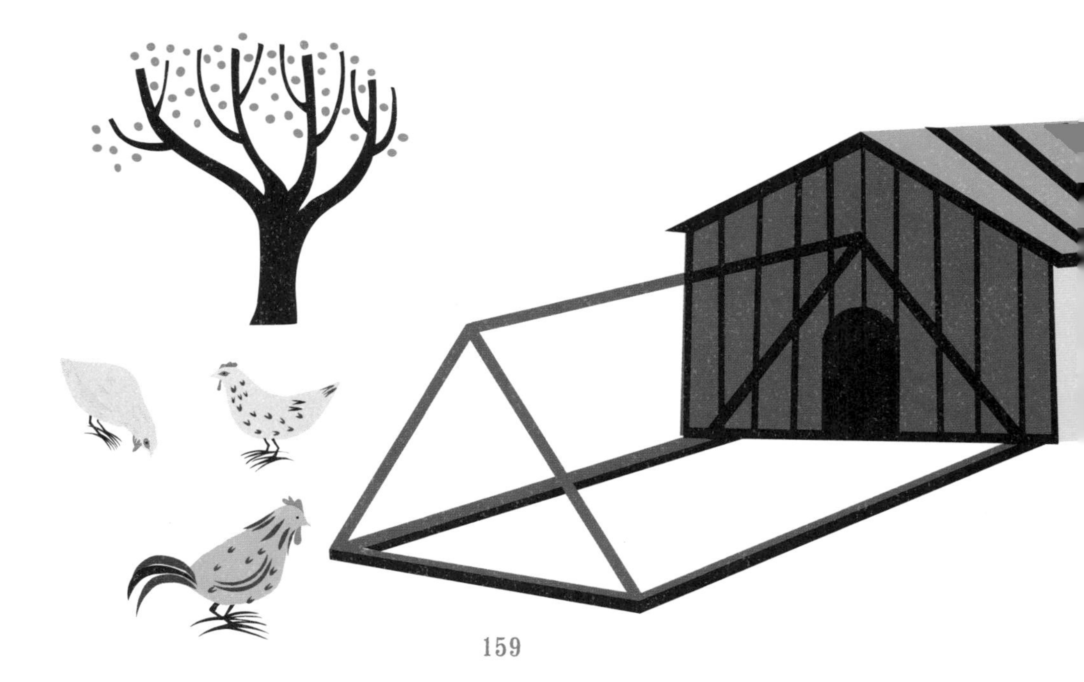

TIP 279: *Consider building a chicken obstacle course!*

It doesn't take very much to keep hens amused, but it's important that you make the effort to do so, nonetheless. One thing you can play on is the love these birds have of getting to a high vantage point from which to survey their domain. Don't construct a scale model of the Empire State Building, or anything grand like that; all you need is to nail together a few lengths of wood, or an old pallet or two, to create a stable climbing frame that'll be interesting for them to clamber over. You may even want to add a sheltered sun terrace at some convenient spot on the structure!

TIP 280: *Try out a "scratch shed" too*

If you've got a bit of space to spare within the main chicken enclosure, you could add what chicken keepers of old would have called a "shed." This is another structure that's separate from the main hen coop, but that offers shelter and shade and overhead cover for the birds to use during daylight hours. It has been suggested that laying hens prefer an alternative form of shelter during the day, so that they can reserve their main coop exclusively for laying and for roosting in after dark. So offering an alternative—it can be a shed (with roosts but no nest boxes) or just a simple, three-sided and roofed weather shelter—can be a great thing to do.

ENSURING A BALANCED DIET

TIP 281: *The right food makes all the difference*

In production terms, hens are what they eat. If you feed them a properly balanced and nutritionally complete layer ration then, all else being equal, they'll respond by delivering the kind of growth and laying performance that's expected of them. However, shortfalls in diet—even quite minor deficiencies in important vitamins or minerals—can have a dramatic effect on productivity and, in some cases, on overall health and welfare levels too. So, while there's probably no pressing need to buy the very best chicken feed you can find, opting for a good, solid midrange product from an established producer makes a great deal of sense.

TIP 282: *Different ages of chicken require different feed*

The image of Ma and Pa Kettle scattering corn for an ever-eager farmyard flock of laying hens on their ramshackle farm is certainly an appealing one. However, poultry keeping's moved on a good deal from this film series, inspired by the book *The Egg and I* by Betty MacDonald. Nowadays, if you want your hens to be productive—whether they are pure breeds or hybrid layers—they'll need to be fed a well-balanced and nutritional diet in the correct form, relative to their age. It's important that you match the feed you buy to the age of the birds you're feeding; chick starter should be given from hatching to six weeks, grower ration from six to sixteen or seventeen weeks, and layer pellets or mash from then on.

TIP 283: Backyard free ranging is good for your hens

Over the years various studies have endeavored to pinpoint just what proportion of a free-range hen's dietary requirement can be met by the food it gathers while foraging. Unfortunately, there doesn't appear to be a definitive answer. One thing you can say, though, is that much depends on the quality of the land the birds are free ranging over. In days of yore, farmers would turn their laying flocks out on to the stubble fields, where the birds would gorge themselves on nature's bounty, both insect and seed based. Nowadays, though, it seems that efficient combines and the extensive use of chemicals in our fields has slashed the rich pickings to virtually nothing. Rather ironically, then, hens are probably better off scratching around in the average urban yard than they would be in a modern field of stubble.

TIP 284: Don't skimp on feed

While your birds will freely eat fruits, veggies, table scraps, and kitchen waste, their primary source of nutrition should be the layer ration. This feed is properly balanced with proteins, vitamins, and minerals, as well as essential amino acids, all for healthy egg production. A "homemade" diet may be deficient in necessary nutrition. Do not feed carb-heavy, dense foods like pasta, bread, and cheese. Your birds will enjoy fruits and vegetables like watermelon and canteloupe rinds, strawberries, tomatoes, lettuce, and blueberries.

TIP 285: *Confined hens will need help with "mastication"*

It's important to appreciate that hens that live their lives in close confinement, and never get access to fresh ground for free-range purposes, must have access to nonsoluble grit. Because chickens have no teeth and so can't chew, they rely on grit that's held in the muscular gizzard (a hollow organ in the intestinal tract, where it's used to grind and break down everything that's swallowed. Under normal circumstances, hens pick up this sort of grit as they peck in the soil, but those without regular access to soil can't. Different ages of bird require different sizes of grit. You may choose to introduce chick-sized grit about three days after hatching, by sprinkling a small amount on the feed. Switch to "grower size" at 4–5 weeks, then to "layer size" from around 15 weeks onwards.

Digestive tract of a chicken

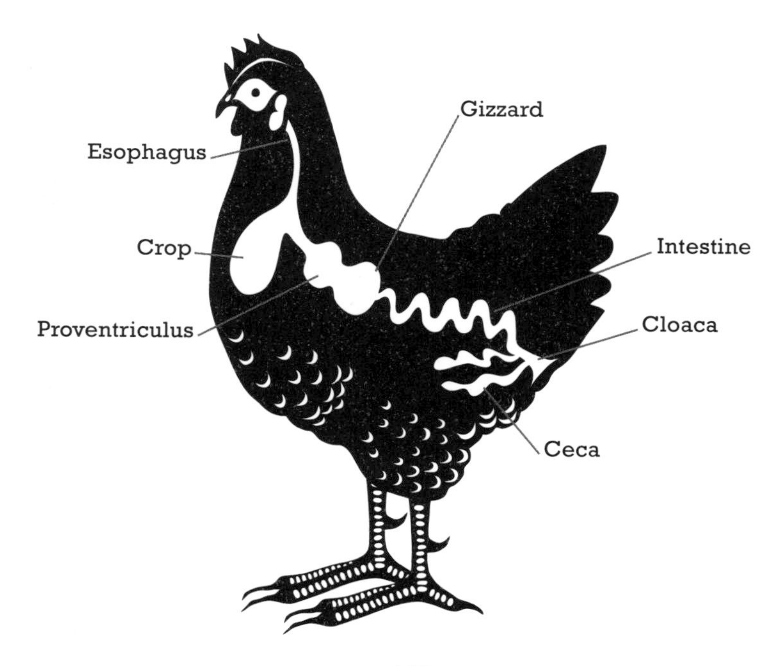

TIP 286: Feeding mash instead of pellets has consequences

Grower and layer feed is available as mash or pellets, but which is best? Well, both contain a mixture of cereal grains, vegetable and animal proteins plus vitamin and mineral additives. In mash, the constituents are ground down to form a dry, powdery mix, while pellets are mash that's been compressed and extruded into small, solid, cylindrical pieces. Mash is the more traditional option, and some keepers find it helpful to mix it with warm water during cold weather, to make it more palatable for the birds. But pellet users argue that there's a lot of waste with mash, as the birds tend to pick out their favorite bits, and also scratch a lot into the ground around the feeders. Using pellets makes selective feeding impossible; every bird gets the full benefit of the feed mix. Also, many users find that pellets work better through gravity-fed feeders; mash can have a tendency to clog, blocking the supply.

TIP 287: *The jury's still out on the need for supplements*

The debate has wrangled on for years about the real-world worth of dietary supplements for chickens. The manufacturers of the plethora of products that now exist to tempt keepers maintain there's a distinct need—well, they would, wouldn't they? Opinions among owners vary, with some saying there's no need at all for any of them, and others believing that the selective use of some is a good thing. The truth, as is so often the case, probably lies somewhere in between the extremes of people's views. Products like cod liver oil, apple cider vinegar, and poultry tonics can all play a part in good husbandry, especially to help bolster birds through times of trouble: during the molt, periods of ill health or after a traumatic experience, or change of scenery, for example.

TIP 288: *Not all treats are good to feed*

There's a lot of common sense associated with keeping a few hens in the backyard—or, rather, there should be. Unfortunately, there are still plenty of keepers (and not all beginners, either) who don't show much sign of having any! The problem is that chickens will eat almost anything with enthusiasm, as though it's the best thing since sliced bread, even when it is sliced bread! So they give the impression they're enjoying themselves, which often has the effect of encouraging their owners to give them even more. Now, when what's being given is fresh fruit or green vegetables, then all is well and good. However, the feeding of fruit cake, pasta, peanuts, chocolate, or bread and jelly is a very different matter. The most traditional "treat" is mixed corn, which, like fruit and veggies, is fine for the birds when given in moderation. Excesses, though, together with the feeding of the sort of unsuitable foods just mentioned, are actively bad for the birds.

TIP 289: Think "natural" when searching for poultry treats

The best advice, if you're determined to give your birds treats, is to stick to foods that you know will do them no harm at all. Natural, unprocessed vegetables and fruits really are the best and safest options. Your hens will go crazy for grapes and corn on the cob, for example, but, certainly with the latter, take care not to overfeed. The last thing you want, when giving treats—even natural ones—is to end up feeding so much that you start reducing the birds' consumption of their normal, proper poultry ration. Substituting a balanced, specialist chicken food with just a single, nutritionally limited alternative is a very bad idea.

TIP 290: Be on your guard for hidden sugar and salt

The key rule, when it comes to the feeding of anything other than properly formulated poultry food to chickens, is to avoid giving them salt and refined sugar. These are both things that they wouldn't normally come into contact with in the wild and, therefore, aren't suited to their digestive systems. Salted potato chips and nuts, jelly, cakes, cookies, and all other forms of confectionery should therefore be on your "Never feed to chickens" list. Nevertheless, people still do it because the birds appear to enjoy them. It really is best to keep their diet as simple as possible, avoiding anything that you think may have been processed in any way. You're doing them no favors by giving them a taste for "exotic" foods, which, however you choose to view it, simply isn't natural for them to be eating.

REASONABLE EXPECTATIONS

TIP 291: *Hybrid hens should be top of the charts*

The careful selection that goes into the creation of today's hybrid laying hens is what guarantees keepers of these birds a virtually unparalleled laying performance. Any hen that can produce 320-plus good-sized eggs in a year is a pretty remarkable creature. However, you'll notice that I included the words "virtually unparalleled" in the first sentence of this tip, because it's not quite as categorical as that. You see, hybrids are absolutely fantastic layers, but only for a short time. In fact, they start slowing down in just their second year of life, which is why commercial operators move them on after only eighteen months of service. So, while a hare-like hybrid will streak into an early lead in the egg numbers race, the pure breeds—in true tortoise fashion—can catch up and eventually win.

TIP 292: *Pure breeds may still represent the best bet*

Although hybrid hens are desired for their excellent laying, good, utility strains of our most prolific pure breeds, such as the Plymouth Rock and Leghorn, can be equally attractive in different ways. Pure breeds cost more to buy but retain their value better. You can breed from them and sell the offspring for decent money. Generally they'll live longer than hybrids, and they certainly keep laying for longer too. Then, of course, you have the heritage and conservation aspects, which are important to an increasing number of keepers. You're literally buying a piece of history when you get yourself some good pure breeds, and productive history at that.

TIP 293: *Rescued hens aren't always the best option*

Few would argue against the well-meaning aspirations of those who devote so much time to the rescue and subsequent rehoming of commercially farmed laying hens. These are the birds that have served their time in an intensive cage-based or barn egg production system, and that are typically slaughtered to make way for the next batch following on to replace them. Rather than see these relatively young birds slaughtered, people choose to offer them a relaxed retirement instead. However, if you want to do something similar, it's important to appreciate that it's not all sweetness and light. These birds generally arrive in a pretty threadbare state, they find the transition from farm to yard traumatic, some are ill or die from the shock and others will never lay another egg. So think carefully before venturing down this ownership route.

TIP 294: *Decide whether little or large is best for you*

Poultry keeping is one of those hobbies that gets people really excited almost immediately and, as a consequence, it's all too easy to get carried away. So novices frequently overestimate their own facilities, underestimate the work involved, and commit to the wrong type of bird. These mistakes, of course, are only fully appreciated a little way down the line, once the chickens come home to roost, you might say. The logical way to proceed is to start small but, then again, people rushing excitedly into a new hobby are rarely logical. Still, beginning with bantams really is the smart way to get your first experience. These little birds (available in just about every pure breed) are easier to keep than large fowl, need less space, eat less food, cost less to buy, look just as pretty, lay just as many eggs (albeit only they are a bit smaller), and are much easier to handle.

TIP 295: *Be sure you're getting the age of hen you pay for*

For many poultry beginners, the obvious time to buy hens is when they're just about to start laying. Known, logically enough, as "point of lay" (POL), this important landmark in a hen's life arrives when it matures, normally at about twenty-four weeks old. Unfortunately, some breeders exploit this slight vagueness as a money-making opportunity, playing on the fact that it's all but impossible to tell the age of a hen just from its appearance. So they sell birds as "POL hens" when, in fact, they're a good few weeks younger than they should be. The seller then gets the money for the early sale and transfers the on-going cost of feeding the bird to the new owner. This is annoying and frustrating for the new keeper, who is understandably anxious for the first egg to arrive. This event could be weeks or even months away. So do what you can to get an age assurance before you buy.

TIP 296: *Plan your options for when the egg supply ends*

There always comes a time when a hen finally stops laying, at which point you're faced with a bit of a dilemma. Much, of course, depends on how you regard your chickens: are they pets, are they livestock or, as is the case with many keepers, are they somewhere in between? Hybrid hens will stop laying first, typically after two or three years. Pure breeds, however, will go on for twice as long, although the total number of eggs will gradually diminish as the years pass. Thereafter you have to decide whether to keep the birds until they die naturally, or whether to save the money you'd spend on feeding them and hasten their end with humane slaughter.

TIP 297: Don't expect to make vast profits

While we all like to imagine that keeping a few hens at home represents a cheap source of fresh eggs, the reality is never quite so optimistic. The impression given is that hens produce eggs for free; they just do it, every day or so, as part of being alive. But, of course, there's always a cost attached. Those involved in home-based hobbies very rarely take into account their own time and, indeed, why should they? But it is a factor if you're seeking to put a price on the eggs being produced by your hens. The setup costs are often quite expensive: the coop and run alone can typically account for $1,500 right away. Added to this, there's the cost of the other hardware needed, plus the birds themselves and, of course, the feed. Then, factor in a few visits to the vet and other necessary additions such as a new wheelbarrow, garden storage box, spare fence battery, utility wagon, and so on, and the total spent in the first year can easily push up towards $2,000. That buys a whole lot of eggs from the supermarket!

TIP 298: Eggs aren't the only benefit of keeping hens

Living with hens in your backyard isn't something you do to balance the books or save money. It's much more of a lifestyle choice, something you get involved in for the greater good. Chickens are fascinating creatures that offer their keepers hours of pleasurable involvement and a wonderful distraction from the hustle and bustle of everyday life. Time spent with your hens should be seen as relaxation, even when it involves a bit of work. It's a time to lose yourself in their world and to forget the pressures of your own. Many keepers talk about the therapeutic benefits of keeping a few hens and, with hindsight, plenty also regard their decision to get involved as being a truly positive, life-changing moment. Now, that can't be bad!

TIP 299: *Sometimes, tough decisions have to be made*

There are bound to be times when, for whatever reason, you're faced with a chicken-related emergency situation. Birds that are seriously ill, or get badly injured, will need to be put out of their misery: remember that as the owner of an animal, you should seek to minimize suffering. So, if it's not possible to get the bird to a vet, or have one visit you within a reasonable timeframe, you'll have to make what most understandably regard as a dreaded decision: to kill the bird yourself. This must be done humanely. (See also tips 437–442.)

TIP 300: *Not all birds will be oven ready*

Eating your own chickens is a tricky issue and many domestic keepers couldn't imagine anything worse: killing and eating the family pet is an understandably upsetting idea. There is an increasing number of owners starting to rear birds for that specific purpose, but doing this requires a particular mindset in terms of attitudes towards the birds. Everybody involved needs to be clear that they are "livestock" rather than pets. But it's important to appreciate that not all birds make good eating. You'll need a recognized, dual-purpose pure breed male that's killed at the right age to make a decent-sized roasting bird, or a modern, broiler-type hybrid. Hybrid layers generally don't develop enough meat to make a worthwhile table bird, and birds that die due to disease are obviously not suitable for consumption.

GOOD HUSBANDRY

As a budding hen keeper, it's very important that you learn how to look after your birds in a way that's going to ensure they enjoy a happy, healthy, and stress-free lifestyle. One vital aspect of this centers on the day-to-day activities associated with looking after the birds. These simple practices—including feeding, watering, egg collecting, coop cleaning, and rudimentary health checks—are collectively referred to as "husbandry." Establishing a good husbandry routine is a basic but fundamental part of any chicken-keeping project.

EVERYDAY CARE AND ATTENTION

TIP 301: Let the birds out on time

Chickens thrive on routine and, once established, they like to maintain regular timings for everyday activities like feeding, laying, and being let out of the hen coop in the morning. So it's important that you fix on a time to open the pop-hole that's both convenient for you and that you'll be able to stick to day in, day out. Naturally, of course, hens would rise with the sun, getting out and about as soon as they found it light enough to do so. Now, while such early starts are not convenient for most keepers, it's important not to keep your birds locked in their coop for too long, once the sun is up. One convenient solution can be a light-activated, automatic pop-hole opener, although a device like this will need to be set accurately, and will usually also require the fitting of a lightweight, metal, pop-hole door.

TIP 302: Examine the feeders for blockages

You'll also need to check the feeders every morning, to make sure that there's sufficient feed inside to last the birds for the coming day, and that it's flowing correctly. Although many gravity-fed units nowadays are fitted with "weather caps" that are designed to shelter the exposed pellets in the feeding ring from rain and falling debris, these aren't always completely reliable. Rain blowing on the wind will have little difficulty in reaching the feed and, once damp, the pellets will start to stick together and clog up the supply. It's important to deal with this sort of problem sooner rather than later because, if left, the dampness will simply spread to spoil and unnecessarily waste more pellets.

TIP 303: *Drinking water is a vital requirement*

It's vital that all hen keepers appreciate just how essential water is to their birds. While domestic chickens (especially free-range birds) can survive the absence of food for a day without too much of a problem, a similar lack of water can pose a real problem. To illustrate the point, a bird can lose all of its fat and recover, or half of its protein and recover. However, if it loses just 10% of its body moisture, it'll die. So dehydration, and its debilitating and dangerous consequences, must be avoided at all costs, and that means maintaining a constant supply of clean, fresh drinking water. Check and replenish waterers every morning, and remove any debris that may have built up during the previous day or overnight.

TIP 304: *Check your hens every morning*

Once the birds are out, spend a few moments just watching your birds as they stretch, flap, and scratch their way into a new day. Visually check each one in turn, assessing its behavior and general attitude. Note any that are slow or reluctant to leave the coop, or that appear hunched or lethargic. All should rush enthusiastically to the feeder or waterer to begin breakfast and they'll all usually produce a large dropping pretty quickly too. If you get the chance to see this happen, check that it looks as it should (firm, with distinct white and dark brown/gray portions). This whole process need only take you two or three minutes per coop, and is time very well spent. Make a mental note of any birds displaying unusual behavior, or producing abnormal droppings, and isolate any that appear obviously ill.

TIP 305: *Always collect the eggs early*

Once you've let out the birds, given them the once-over, and checked/ replenished their feeders and waterers, you need to check the nest boxes (and floor of the coop) for eggs. As a general rule it's important that eggs are collected as soon after laying as possible; leaving them in place simply increases the risk of them getting cracked or even broken as subsequent birds use the nest box. Broken eggs inside the coop are really bad news because it's very likely that the hens will start eating the contents. Once this happens, and they realize how much they like it, it can quickly turn into a habit, and they'll start breaking into other eggs simply to eat them. An egg eating habit is a tough one to cure once it's become established, so don't give your birds the chance to get started in the first place.

TIP 306: *Don't forget the fence*

Finally, as part of your early-morning session with the hens, visually check or, if your enclosure is larger, take a walk around the perimeter, to make sure that all's well with the fence. The hours of darkness, or the period immediately after dawn, is when predators are most likely to strike, especially in rural areas. So if damage has been done by digging foxes or coyotes, or even gnawing rats, then it's first thing in the morning when you're most likely to notice it, and it gives you the rest of the day to sort it out. Also make sure, if you're using electrified netting or wires, that these are functioning properly, and that you turn them on again after leaving the pen. It's all too easy to leave without doing this; a mistake that can be very costly for those whose birds are unattended during the day.

TIP 307: *Make time for a lunchtime inspection*

If you're able to do so, it's a very good idea to spare a few minutes in the middle of the day to visit your birds for a general inspection, and to collect any further eggs that may have been laid during the morning. After a bit of time with your hens, you get to learn which birds lay when, and you'll be able to time your visits to the pen accordingly. Regular inspections are always a good idea. Not only are they the quickest way to get birds used to and confident in your presence, but it's good husbandry to keep an eye on everything as frequently as you can. Trouble can strike very quickly in the poultry pen, and things can turn from minor to major in a very short time. So those keepers regularly on hand will be best placed to nip problems in the bud.

TIP 308: *If it's cold, give a mixed-corn scratch feed*

Chickens are generally excellent at controlling their own body temperature—during both hot and cold weather. While some of the smallest true bantams, such as the Belgian and Sebright, might not relish being outside in the bitterest of winter weather, most are resilient enough to get on with life pretty much whatever the weather. However, one thing that keepers can do during cold snaps is to give their birds a handful or two of mixed corn. For the best results, scatter this on the run floor so that they have to scratch around to find it. Not only do they love eating mixed corn, but the activity of doing this will work in their favor and the energy released from a crop full of it will certainly help them pass a more comfortable night when temperatures are at their lowest.

TIP 309: *Never forget to close the pop-hole*

Another quite common mistake made by keepers, especially those new to the hobby, is to forget to close and secure the hen coop pop-hole after dark. Some people, of course, don't forget, they just can't be bothered because the weather is bad or they're engrossed in a movie. Whatever the reason, though, not securing the hen coop is a risky business that exposes your birds to an unnecessary level of danger. Never forget that it's an important part of your duty of care towards the birds—as a responsible keeper—to protect them from all common dangers, to the best of your ability. Failing to take this basic security measure is letting them down badly.

TIP 310: *Empty waterers if it's going to freeze overnight*

This may seem like an obvious precaution to take, but plenty of keepers don't bother, and regret it in the morning. For the sake of the few seconds that it takes to upturn a waterer or two when you shut the birds in for the night, you can save considerable bother the following morning. Time is always short at breakfast time, with family members rushing to get ready for school, college, or work. So you need your morning session with the hens to run as smoothly as possible; dealing with a waterer that's frozen solid (and maybe even cracked as a result) can be a real headache. So empty them at night and, just to be on the safe side, keep a can of water in your coop overnight so that you've got a convenient and unfrozen source to refill from in the morning.

EFFECTIVE CLEANING

TIP 311: *Understand the need for cleanliness*

Although chickens are remarkably resilient creatures, their good health remains very much in the hands of their keepers. Those who fall short with their husbandry routines will more than likely create a health challenge for their birds that no amount of natural resilience can match. One of the most basic requirements is to keep the environment in which the birds live as clean and parasite-free as possible. This means both outside in the run as well as inside in the hen coop. Ground that's allowed to become packed down, due to overstocking with birds, loss of the vegetative cover, and contamination of the surface with droppings, will present a health risk to the birds, as will the inside of a hen coop where the litter has become moldy and fetid.

TIP 312: *Little and often is the best approach*

In terms of keeping both your hen coop and the area of ground around it in good condition for your birds, the best advice is to tackle the cleaning operations "little and often." It's far easier to keep everything under control if you're staying on top of potential problems, rather than letting them become established, then trying to deal with them. Lots of people find excuses to put off cleaning out the hen coop, assuming all will be well and, while in most cases it probably will, there will certainly be times when it won't. If ever you're in any doubt, just put yourself in the position of one of your birds. Would you like being cooped up in the hen coop at night, or would you want the smelly, messy old bedding to be replaced before you set foot in there?

TIP 313: *The hen-coop-cleaning basics are easy*

There's no great art to keeping a hen coop clean; it's not a skill that's going to take you years of repeated practice to master. Nevertheless, despite the relative simplicity of the task, too many keepers fail to keep up with this most basic of husbandry jobs. At the most fundamental level, the primary objective in the maintenance of your hen coop environment is to keep the bedding material fresh and clean. Dampness and unpleasant odors (typically that acrid, ammonia smell) are the things to be avoided. A good hen coop shouldn't smell of anything. Happy, healthy hens in a well-maintained enclosure shouldn't smell of anything. The idea that all chickens smell is completely wrong; only badly looked after, poorly treated hens will smell.

TIP 314: *Go that extra mile when cleaning*

As with all cleaning-related jobs, the ultimate success of clearing out a hen coop will depend on thoroughness and attention to detail. Take your time removing the old coop litter, ensuring that you remove those awkward clumps of material that always seem to get stuck in the corners. If the wooden floor has become damp in places, ask yourself why. Is it being caused by rain water getting in from outside, or is it simply that the inadequate layer of litter material above the patch had become contaminated and sodden? Maybe an egg had been broken there... Whatever the reason, take appropriate action; don't just cover it straight back up again. Fix the leak if there is one, and give the dampness time to dry with all the doors open. Use a sanitizing powder to dry and freshen what's left before laying down a generously thick (4in / 10cm) layer of new material.

TIP 315: *Remember to pay attention to the poultry waterers*

The simplicity of the typical poultry waterer means that keepers often inadvertently ignore them, other than to check that they contain water every now and then. Such oversight can cause problems, not only will the birds themselves be responsible for the gradual contamination of the waterer being used (everyday use leads to the buildup of greasy, slimy deposits), but the inside of the vessel itself can host all sorts of bacteria and mold growth. The heating effect of the sun means that the microenvironment inside a half-filled waterer remains warm and moist for much of the time; ideal conditions for the development of trouble. Green-colored algae is a common sight on the inside of badly maintained waterers, and should be removed with an appropriate and safe cleaning product.

TIP 316: *Always make sure chicken feed stays dry*

With the price of chicken feed becoming ever more expensive, none of us can afford to be wasting it. The chickens themselves will be responsible for a fair degree of loss (they inadvertently flick food out as they peck), so it's important that we do all we can to limit other needless losses. Pellets and crumbles can soak up water like a sponge. The result is a sticky, soggy mass that then dries and hardens into deposits that the birds won't like eating, and that will inhibit the flow of feed around them. Even the best feeders can be susceptible to driving rain, so the only way to eliminate this risk is to put them under cover. Don't site them inside the hen coop; instead, build a simple feeding station nearby, perhaps with three sides and a roof, to shelter the feeder from all weather-related problems.

TIP 317: *Don't forget to clear the feed bin*

If, like many hen keepers, you sensibly store your poultry feed loose in a rodent-proof metal trashcan, then be aware that, every few weeks, it's important to clean this out too. The act of using and refilling the contents over a period of weeks leads, inevitably, to the buildup of dust and other debris at the bottom of the container. This residue, of course, is simply a powdered version of the feed, and therefore contains the fats and oils that give poultry feed its shelf life. Remnants left for any length of time at the base of a feed bin will start to turn rancid, leading to the possibility of the good feed on top becoming tainted as well as the development of mold and spores. All this is easily avoided, though, by simply remembering to empty out the bin every few weeks.

TIP 318: *Dealing with dirty eggs*

It's almost inevitable that, every now and then, your hens are going to produce the odd dirty egg. These can be smeared with residue from the laying hen itself, or be contaminated by droppings in the nest box. Either way, it doesn't look good. However, washing them can pose problems, due to the porous nature of eggshell. The risk is that the temperature differences set up between the inside and the outside of the egg during washing can cause bacteria to be drawn through the shell to contaminate the contents within. So, if you can get away without having to wash, that's the best approach. If it's necessary, though, because you're passing eggs on to a neighbor, then use a specialty egg wash product, water that's warmer than the egg, and do it at the last possible moment before parting with the eggs.

TIP 319: *Keep your garden storage box spick and span*

Don't overlook the fact, if you've built a garden storage box to use as a handy repository for all your poultry-related paraphernalia, that the chances are that spilled feed and bedding material will build up in here too. The presence of these materials, scattered over the floor as they'll inevitably be, will attract the unwanted attention of rodents. Squirrels, mice, and rats will be drawn to this easy source of food and nest-building materials and, in getting in to "harvest" the rich pickings, they'll more than likely cause some structural damage. This, in turn, could lead to water leaks and accelerated wood rot. The answer is to keep the interior of the garden storage box swept clean and peppered with mousetraps at all times.

TIP 320: *Be aware of cross-contamination*

If you're living the good life and looking to produce as much of your own food as possible, you'll probably be considering or already keeping other types of livestock. Rearing turkeys, pigs, and sheep for the freezer is all perfectly possible with good organization and an acre or so of land to play with. However, if you go down this route, it's important to realize that you need to take care to avoid carrying problems from one enclosure to another. Turkeys, in particular, should be a particular concern, as they are vulnerable to a disease called blackhead. While chickens can be vulnerable too, it's much more common in turkeys. So, because the disease is spread by a small parasite that gets expelled from the carrier's body in egg form in droppings, it's important not to tread contaminated droppings from the chicken to the turkey pen on dirty boots. When doing your rounds, always visit the turkeys first.

PRACTICAL ISSUES WITH YOUR BIRDS

TIP 321: *Handle your hens regularly to assess weight*

It's always useful to handle your birds on a regular basis. It's important that they be comfortable in your presence, and that they trust you. Interacting with them like this regularly will help build that bond between you and them. As well as being a vital way of checking for parasite and small injury problems, the regular handling of your hens will allow you to remain directly in touch with their general condition and body weight. As your ability to assess their body weight grows, you'll develop the useful skill of being able to detect if the bird is starting to get thin for some reason. Apart from feeling lighter overall, you'll also notice that the breast bone starts to feel "sharper" against your hand. Chickens generally only lose weight for bad reasons, so never ignore this trend.

TIP 322: *While handling, check leg scales for smoothness*

While handling your hens, don't forget to check the condition of their legs, in particular, the smoothness of the scales. These should be tight and close-fitting on a healthy bird, and those that aren't should set the alarm bells ringing with regard to an unpleasant condition called scaly leg. Caused by the scaly leg mite, this problem should be pretty easy to spot, as the edges of the leg scales start to be lifted as the mites begin burrowing underneath them to find refuge, food, and to breed. As the condition develops, the scales lift further and whitish-gray, crusty deposits start appearing around the scales. By this point the bird will already be in significant discomfort; it's an incredibly irritating and painful condition. Despite the pretty obvious signs, though, many keepers fail to spot the problem quickly enough, so please be aware of this possibility.

Body parts of a rooster

TIP 323: *Listen for noises that may indicate health issues*

Chickens can suffer with "summer colds," when they present wet nostrils and watery eyes, sneezing, gurgling, wheezing, and other respiratory noises. The possible causes are many and varied, but some of the most popular include over crowding, inadequate hen coop ventilation, poor nutrition and bad feeding. However, while there's no doubt that these factors can certainly contribute to general health, the latest thinking would suggest that the cause of summer colds in hens is likely to be a respiratory virus, and that the timing is coincidental. One other more serious possibility with similar initial symptoms is infectious bronchitis, which is a chicken-only virus.

TIP 324: *Don't let their feet get toe-nail balls*

Birds that are forced to live in muddy conditions will find it a challenging experience, and not in a good way. A generally wet environment is far from ideal for chickens, and will greatly increase the amount of input needed from the keeper to ensure the coop itself remains habitable. But the birds can suffer in a more direct way too, when the outside conditions are too wet. I've seen examples of a bird's feet that had become so caked in mud that each toe was tipped with a ball of solid mud, about a ½in (1 cm) in diameter. These mud balls had become so hard that they had to be very carefully broken away using pliers. As you might imagine, the bird had become extremely lame and was in considerable discomfort. It was extremely careless of the keeper to have let the bird fall into this uncomfortable state.

TIP 325: *Treat your birds to a regular pedicure*

As chickens spend so much time on their feet, and are always scratching at the ground in search of tidbits, it's very important that their feet, toes, and toenails be in good shape. The underside of the feet can suffer with cuts and splinters that, if they turn septic, will be both painful and debilitating. Check regularly and seek veterinary help if you discover this sort of problem. Toenails grow like ours do and need to be kept at a comfortable length. Overgrown nails can be a problem for birds that don't get outside much, as the natural act of scratching hard ground won't be helping to keep length in check. They need to be clipped regularly (only the light-colored tip) to prevent discomfort and possible toe injury. Don't overshorten otherwise they'll bleed.

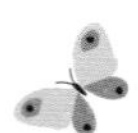

TIP 326: *Always pay attention to a hen's eyes and nostrils*

You've probably heard the saying "Eyes are the window to the soul" and, whether or not you believe that may be applicable to chickens, their eyes can certainly provide a very useful indicator about overall health. So, when handling your birds as part of your regular good husbandry routine, make sure that you take a close look at the condition of their eyes. Any watery discharge (especially when combined with wet nostrils) can be regarded as a warning sign that the bird may have a respiratory problem, such as the very contagious mycoplasma. Sufferers should be isolated and veterinary advice sought. Eyes that appear small or sunken can be a sign of dehydration, or that there's something more serious wrong.

TIP 327: *Large combs can be vulnerable in winter*

If you keep a breed of chicken with a large, single-type comb, such as the Leghorn, the Minorca, or the Spanish, then be aware that these (and their wattles) can be vulnerable to frostbite during spells of cold winter weather. Frostbitten parts turn bluish-red in color, become swollen, and are painful. The tips will turn black, gangrenous, and eventually fall off. If discovered early enough, the frosted areas can be thawed out using cold water, after which they should be dressed with Vaseline; the bird must then be isolated during convalescence. If badly swollen, the damaged parts may have to be removed to ease the suffering and aid recovery; this is a specialist job for a vet. Problems with frostbite can be guarded against by applying a good layer of Vaseline®.

TIP 328: *Keep beaks trimmed at all times*

Chickens' beaks grow constantly (like their toenails), and need to be kept at an acceptable length if eating and beak-splitting problems are to be avoided. The upper beak normally grows quickest, curling downwards over and beyond the tip of the lower one as it does so. An overhang of more than about ¼ in (5mm) is likely to start making it difficult for the bird to peck effectively and pick up food. Unfortunately, though, it's quite common for novice keepers not to notice (or be aware of) this problem. When trimming a beak, the aim is to remove most of the lighter-colored, excess growth, as you might with your own fingernails. Cut straight across, then carefully and gently trim the square corners so that the beak starts to develop a pointed tip. Use a nail file to carefully round off any sharp edges to create a rounded tip and remove any jagged edges that might lead to splitting.

Potential beak issues

Crossed beak

Open beak

Wing of a chicken

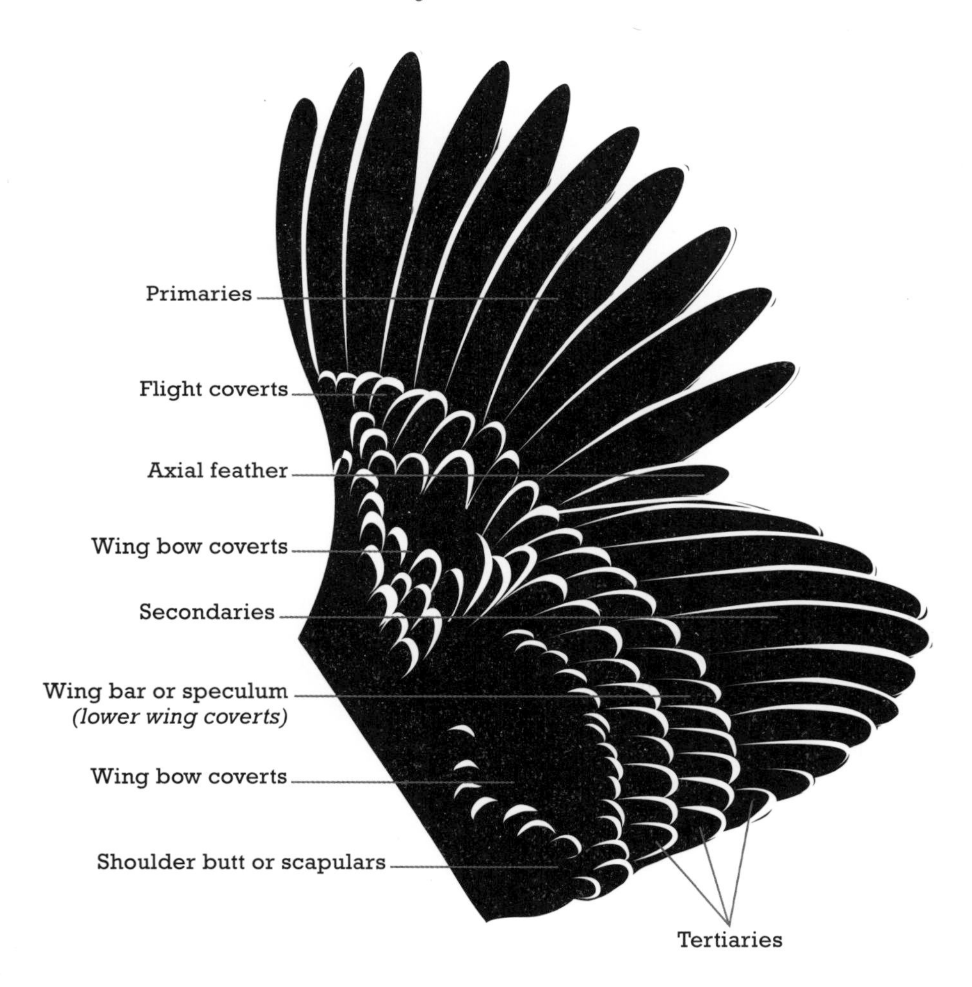

TIP 329: *Dirty feathers should never be ignored*

From time to time all hens will produce looser-than-normal droppings, which can have a couple of noteworthy effects as well as indicating potential problems. At a practical level, they can cause an unsightly and sticky mess in the feathers around the vent, which, as it builds up (hens defecate many times a day) will start to contaminate the eggs as they pass. This sort of buildup will also start to attract flies and other undesirables, with the obvious health threat that that poses. In bad cases, the dirty feathers will need to be washed, or cut away. But it's also important that you think about what's caused the watery, loose droppings in the first place, especially if they continue for any length of time. Possibilities include disease, parasitic worms, lack of food, too much salt or potassium in the ration, sudden changes in environment or husbandry, cessation of lay, and digestive upsets.

TIP 330: *Wing feather tips will need reclipping annually*

Don't forget that if you have containment issues with your birds (bantams and light breeds are typically better fliers than the heavy breeds), then any wing clipping that you do to keep them in their enclosure will need to be repeated annually. Every time the bird molts and replaces its wing feathers, the tips of the new ones will need to be removed again. Remember that you only need to take the first 2in (5cm) off the primary flight feathers of one wing. The idea is to throw the bird off balance so that controlled flight is impossible. Take care not to cut into blood feathers, which are found generally during a molt. Cutting into a blood feather may cause blood loss. You can spot bloodfeathers because their quill will be dark with blood supply. Once the feather is fully grown, the quill becomes translucent. As a general rule, most keepers opt to cut the feathers on the left wing.

PARASITE-FREE HOUSING

TIP 331: *Be sure you're aware of the potential problems*

Along with lice, red mites are a common parasite to affect chicken coops nowadays. It's very important to be aware of the problems this pest can cause, both for your birds and for you. The dramatic speed with which it reproduces and matures means it can take as little as 10 days to complete the egg-adult-egg cycle, so infestation can develop at an alarming rate. The consequence of letting these tiny creatures colonize your hen coop unhindered is that eventually they'll start discouraging the birds from roosting in there at night. The blood sucking that goes on after dark becomes unbearable, so the chickens become understandably reluctant to be shut inside. There's also a very real risk of you carrying red mites in to your house, where they can colonize beds and other soft furnishings.

TIP 332: *Make your antimite measures effective*

If you're determined to provide your hens with the best possible environment inside their coop, then you need to be thoroughly proactive with your anti-red mites measures. Don't wait until you find signs of their presence before acting—use the right products to help prevent them from becoming established in the first place. Consult your local poultry specialist supplier, as well as experienced keepers in your area, to get their advice on the best way to tackle this problem. Most people you talk to will have a different take on what's needed, so it's important to gather as much experience-based information as you can before deciding on your strategy.

TIP 333: *Keep your mite control expectations realistic*

There's a school of thought that suggests that it's all but impossible to eradicate red mites from a wooden hen coop once they've become established inside. The way they live—seeking refuge in all the nooks and crannies of the structure—means that tackling them effectively is a significant problem in most cases. The best that the majority of keepers with this problem can hope for is to keep these pests under control, by limiting their numbers to a level that's not detrimental to the birds. Owners of plastic coops that can be taken completely apart (because they're simply pinned and clipped together) probably represent the best hope for really effective red mites control. The fact that wooden coops can't be stripped down in this way means that most mites will never be exposed to the chemical treatments.

TIP 334: *Always use antipest products as recommended*

A common mistake that plenty of chicken keepers make—both beginners and those with greater experience—is to misuse antimite products. There's an all-too-common tendency for people to use a new product and then to sit back assuming that that will have done the trick. If only it were that simple. The reality of it is that, once you've started the fight against the mite, you can't really stop, ever. To effectively break the life cycle of the red mites, you need to be treating with the appropriate product every week or so, to make sure there's fresh treatment waiting for each new batch of eggs that hatches.

TIP 335: *You don't have to rely totally on chemicals*

It's not all about chemicals, though, as there are some extremely valuable "weapons" available for the fight against mites that are natural in origin. Diatomaceous earth (DE) is a perfect case in point. This very fine, slightly abrasive white powder is made up of the fossilized, skeletal remains of single-celled plants called diatoms, which existed in freshwater lakes some 30 million years ago. This silica-based material is lethal for insect life for two reasons. Its sharp-edged form is sufficient to cut through an insect's exoskeleton, after which the porous nature of the product ensures that fluids are drawn out of the body to cause a rapid death by dehydration. This powder can be puffed into crevices, added to dust baths or applied directly to the birds. However, while DE can be beneficial in the coop, it can also kill helpful microorganisms in the soil.

TIP 336: *Always give powders a chance to work properly*

Most of the approved antimite products available are powders and only work on contact with the target, so they won't do any good if they're not applied effectively. It's important to appreciate that, although you might have given the walls a good dusting on the inside of your hen coop, this will offer no guarantee of success. To be as effective as possible, the powder must be driven as deeply as possible into every joint and crevice that you can see. The coop must be empty, with roosts and nest-box units removed, then thoroughly vacuumed to clear all loose debris from the corners and joints. Then, dust the coop thoroughly with the powder. You should also dust your birds by placing one bird at a time in a garbage bag (with the head sticking out) along with the dust. Hold the bag closed to powder its body.

TIP 337: *Learn to recognize the signs of infestation*

For the benefit of those of you who haven't yet bothered to apply any antimite treatment because you don't believe there's a problem, here are a few of the signs to look out for. Watch out for ash-gray, powdery deposits, typically found close to joints in the structure of the coop or in the roost supports. This is the waste product produced by red mites. After dark on a warmish evening, open the main hen coop door slowly and quietly, just enough to get your arm inside. Run a finger along the underside of the roost, close to where it meets the wall. A smear of red on your fingertip indicates that you've killed some mites, and that you have a problem. Finally, if the hens become reluctant to go into the coop to roost for no apparent reason, then the red mites population within might have reached infestation levels.

TIP 338: *Make your wooden hen coop mite-unfriendly*

So the name of the game is to make your hen coop as unfriendly towards the likes of red mites as you possibly can, by the regular use of effective antimite products, wood preservatives and any adjustments to the interior that you feel may help. Some people opt to paint the insides of their hen coops, believing that this helps to seal up the sort of cracks and joints that mites like to call home. There are some very thick paints on the market nowadays (used in the agricultural sector) that are pretty effective at this, or you can use caulk, although you do need to double check that there's no toxicity issue. The supplier or manufacturer should be happy to advise in this respect.

TIP 339: *Be aware of the restrictions on the use of creosote*

For years and years, poultry keepers routinely applied creosote to their hen coops, as people did to their sheds too. Made from a complex mixture of coal tar derivatives, it proved to be an excellent product for both preserving wood, and controlling the insects and fungi that like to destroy it. As a consequence, it was an effective measure against red mites as well. However, in 2003 the US Environmental Protection Agency decided that the use of creosote should be restricted to "professionals" only. Studies had revealed an increased risk of skin cancer among those in daily contact with the product over a working lifetime, and this was felt serious enough to end its "amateur" use. Nowadays, while the product remains on sale for those involved in commercial pursuits, its availability to the general public is restricted. The EPA continues to work in close cooperation with the Canadian Pesticide Management Regulatory Agency (PMRA).

TIP 340: *If all your best efforts have failed, then...*

The red mite is a tenacious survivor, there's no question about that. Couple this with the fact that, ironically, the average, traditionally-built wooden hen coop offers it just about the ideal living environment, and you can appreciate the potential for trouble. The speed with which red mites breed and develop means that numbers inside a chicken coop can reach infestation levels in just a few weeks. So, if keepers aren't on their guard and fully aware of the need to be proactive against this pest, then the battle can be lost before it's even started. In cases where a coop is riddled with mite, and no amount of treatment will sort things out, owners may have to resort to destroying the structure and starting again with a new one. Burning really is the best way to ensure their eradication, but check your local laws before lighting a match!

BATTLING WITH THE PARASITES

TIP 341: *Understand your creepy-crawly enemies*

Essentially there are two ectoparasites that you'll commonly find living permanently on your hens—lice and Northern fowl mite. Infestations will cause a significant drain on the bird's resources, and will create irritation that can be bad enough to affect egg production. Lice are small, yellowish-gray, six-legged insects. There are some 40 or 50 species that can be found on chickens; different species may be seen on the abdomen, around the vent and tail and on the wings and neck. They aren't blood sucking parasites, but have cutting and biting mouth parts to feed on feather material and scales from the skin. The harm they do is caused by the severe irritation and discomfort that their spiny bodies and sharp claws provoke.

TIP 342: *Lice need to be dealt with quickly and effectively*

Lice spend their entire life on the host, starting as clusters of eggs at the base of the feathers—known as nits—which are often found close to the vent (the warm, moist conditions there suit their development). These parasites can't survive long off the body of their host, even if other food sources are available, so spend their whole lives driving their hosts to distraction with their irritating and unpleasant activities. Treatment involves dusting all birds with DE or a suitable insecticidal antilouse powder, following the manufacturer's recommendations. Allowing the birds to use dust baths (to which the DE or powder can also be added) will also be effective in the control of these pests. Antilice products need to be administered repeatedly: many keepers make the mistake of assuming they are a one-shot treatment.

TIP 343: *Northern fowl mite tend to prefer male chickens*

Northern fowl mite are tiny, dark brown creatures that are typically found around the vent and tail region of a chicken. Unlike red mites, the Northern fowl mite remains on the bird at all times, laying its eggs among the feathers. These pests suck blood, irritate, cause dry, scabby skin and, in so doing, debilitate the bird. The mites prefer to live on the males so, when looking for infestation, it's best to inspect the roosters first. Treatment involves dusting birds with DE or a suitable insecticidal powder, and repeating this every seven days or so, until the problem is cured. Adding the powder to the birds' dust bath is a sensible, additional measure to take. Some mites will inevitably fall off the birds, so it's important to treat the coop as well. They can survive for up to three weeks off the bird, so take time to clean and spray the coop thoroughly. A sprinkling of diatomaceous earth will always help too.

TIP 344: *Worm your birds if you need to*

Some keepers (and vets) take the view that there's too much fuss being made about worming chickens these days, and that much of this is due to the many suppliers pushing for the sale of their various product lines. Recommendations for worming domestic chickens varies from not at all to every six months. The key to controlling parasites, however, is largely management: good husbandry, removing any intermediate hosts, not mixing birds of different ages. All chickens with access to the outside are likely to have some degree of worm burden; it's a natural consequence of the way they live. But this will only become an issue if something else is amiss so, assuming it's not, why worm?

TIP 345: *First of all, confirm you actually have a problem*

Nobody wants to administer unnecessary treatments to their chickens: not only is there the expense to consider, but there are many keepers who feel increasingly uneasy about the regular use of chemical treatments on their birds. While it's sometimes possible to judge whether or not your hens have a problem with parasitic worms—they can become generally listless, disinterested, and laying may be affected—recognizing these signs can be tricky for inexperienced keepers. Fecal analysis represents the only way to be completely sure about the situation. It's not expensive to have done, and your local vet should be happy to make the necessary arrangements. The results will highlight the level of worm burden and should be used as the most authoritative guide about whether to worm or not. Your vet will advise on the best product.

TIP 346: *Check your vet's poultry-related credentials*

It's a sad fact that the majority of vets have very little experience when it comes to dealing with domestic poultry, or birds in general. The problem stems from the fact that chickens simply haven't figured in the veterinary training program to any significant degree. Typically, students studying veterinary science at universities in the US do not receive extensive avian training. However, there are some vets who have avian certification. While this mostly applies to exotic birds, chickens can be treated similarly. It's important that you find one of these if you want meaningful diagnoses of chicken-related problems.

TIP 347: *Take care with the natural approach*

There's an increasing number of "natural" poultry products on the market, which make seemingly wonderful claims regarding the effects they have. They are appealing in many ways, especially to those who rail against the excessive use of chemicals. However, it's important to keep in mind that products based on herbs, vitamins, and minerals, while being unlikely to ever cause any harm to your hens, might not always be as effective as it is claimed by their producers/suppliers. Nobody wants their birds pumped full of chemicals, but lurching wildly to the other extreme can be equally undesirable if you're not careful. Your chickens are totally dependent on the measures you take to keep them healthy. If you fall short, it's their welfare that will suffer, not yours. You have a duty to care for your birds, and are legally obliged to do all within your power to keep them fit and happy.

TIP 348: *Diatomaceous earth is great value for the money*

In these days of escalating feed costs, the expenses associated with keeping hens at home seem to rise ever higher. Antiparasite treatments aren't cheap either and, regrettably, there are plenty out there that simply don't do what they say on the package! Consequently, inexperienced keepers can end up throwing good money away on ineffective products that offer nothing more than a slick marketing campaign. There are exceptions, of course, and for the best impartial advice you should seek out the experienced-based views of members from your breed club or local poultry society. One that I always recommend, simply because there's no doubt that it's effective and it also offers great value for the money, is diatomaceous earth. Economically it's great because it doesn't degrade, wear out or lose its potency. If it's there on the floor of the coop, or in the birds' dust bath, it'll be working.

TIP 349: *Take great care with the use of poisons*

The fact that keeping hens in your backyard can be a bit of a magnet to the local rodent populations means that you need to be extra vigilant with your anti-rat measures. Nobody likes to see rats scuttling around close to their coop, and neighbors will be singularly unimpressed if they spot them and make a connection (rightly or wrongly) with your chickens. For this reason, it's very important to understand and use poison in a responsible and safe way, even if you've seen no evidence of a rodent presence. You can rest assured that they will be out there, waiting to take advantage of your generosity with free food and eggs. Buy properly designed bait stations to ensure that the poison you use remains completely inaccessible to all but the intended recipients. Place them with care and talk to other poultry keepers and specialist suppliers about which poison to use. Finally, always follow the manufacturer's instructions to the letter.

TIP 350: *Check local laws before releasing trapped animals*

If you've set traps and managed to catch a raccoon, squirrel, rat, opossum, or other nuisance animal, your city or region may have laws for relocating or disposing of the animals. Consult you local laws before releasing the animal again.

HATCHING AND REARING

Even the most hardened and experienced poultry fanciers never get tired of the thrill that accompanies the hatching of a fluffy new chick. Whether you've entrusted the job to a broody mother hen, or relied on the use of an incubator, the end result will be just about the cutest thing you can imagine! But wonderful little chicks are just the midway point for chicken breeders: getting them to hatch in the first place is one thing, while rearing them on successfully to adulthood is quite another.

INCUBATOR BASICS

TIP 351: *Choose your incubator carefully*

With so much choice on the market these days, buying an incubator can be a daunting prospect. Egg capacity and egg turning method are two fundamental factors. While large-capacity incubators are great, and often work better than the smaller ones, they also cost more to buy and can tempt you into hatching too many chicks! The most basic units will require you to turn the eggs by hand, each day (replicating what the broody hen does naturally on the nest). Failure to do this will impact significantly on successful hatch rates. Semi- or fully automatic machines will do most or all of the turning work for you. As a rule, it's always best to buy a recognized brand, from a manufacturer offering a decent warranty.

TIP 352: *Beware of the manufacturer's marketing hype*

There's a growing tendency for even quite middle-of-the-road incubators to boast full electronic control. However, don't imagine that this makes them foolproof in terms of hatching success. Maintaining the correct temperature and humidity levels inside the machine, and combining this with effective egg turning, is a balancing act at which, as you might predict, the premium machines are usually better than the more budget-priced models. So don't always believe the marketing hype and assume that just because a new model has lots of flashing lights and buttons, it'll do a better job than a more established but less "state-of-the-art" one. The best advice is usually to opt for a machine that's been recommended to you by an experienced breeder.

TIP 353: *Make sure it's a clean machine*

Eggs must always be incubated in a machine that's thoroughly clean inside. Maintaining an atmosphere within that's both warm and moist provides the perfect breeding environment for bacteria. So it's vital that the unit is properly sanitized before every use. Even brand-new machines should be carefully cleaned and disinfected before the first batch of eggs are "set." Eggshells are permeable, and so easily breached by potentially harmful bacteria that can develop inside the unit during the typical twenty-one-day incubation period. Remove everything that you can from inside before you start washing, and clean these bits separately. Take care to keep moisture off any electrical components or exposed wiring, and always make sure to use a suitable cleaning product that isn't going to damage the parts.

TIP 354: *You can't ask too much of your incubator*

Although modern incubators are great machines, they aren't miracle-workers under all conditions. You must give some thought to the placement of your unit, avoiding extremes of temperature. For the embryo to develop successfully, the egg must be maintained at a constant temperature of 100°F (37.8°C) and, depending on the design, the incubator will use light bulbs, heating elements, or electric fans to achieve this. However, this vital temperature can be maintained reliably only if the incubator is placed somewhere where the ambient temperature remains at 63–73°F (17–23°C). So direct sunshine is a real no-no, as is a freezing cold garage or utility shed. Use your common sense and pick a situation in which the incubator can do its delicate job properly.

TIP 355: *Always check before you start*

However simple the instructions may make the use of an incubator seem, you still need to take time to check that the unit you have is actually working properly before you start the process for real. The incubation of eggs is an extremely precise process, with very little tolerance in terms of temperature and humidity variation. As such, the consistent maintenance of both for prolonged periods is the minimum you should expect from your incubator. To check this, it's important to run your machine for at least 24 hours before loading it with eggs, and to monitor temperature and humidity levels at regular intervals during this test period. Fluctuations should be regarded with suspicion, and any performance problems must be rectified (with the help of the manufacturer or supplier), before you get started with the eggs.

TIP 356: *Never count on a full hatch*

In practice, it's unlikely that all the eggs you put in your incubator will produce healthy chicks. One of the most common causes of poor hatchability—apart from operator error with the incubator itself—is the substandard quality of the hatching eggs being used. With more people now relying on internet auction sites and mail-order suppliers as sources for their eggs, the chances of hatching problems are on the increase. Remember that even the best-quality, fertile eggs, produced by healthy and well-kept parent stock, rarely deliver a 100% hatch. So, factor-in the vagaries of the postal service and "internet breeders," and chick numbers can actually be surprisingly low. To maximize your chances of hatching success, the best thing to do is collect the eggs yourself, from a local breeder who you know and trust.

TIP 357: *Keep an accurate record of the age of eggs*

It's important to appreciate that old hatching eggs certainly won't be as reliable as younger ones in terms of producing good hatch rates. Age, of course, is relative and, in this instance, I'm not talking in months or even weeks here, but days. As a general rule, hatchability starts to fall away quite dramatically once a fertile egg passes the seven-days-old point. While a hatch is still perfectly possible, the odds against it are increased, as is the possibility of something affecting the overall health of the resultant chick.

TIP 358: *Second-hand machines can spell trouble*

Buying cheap, second-hand incubators really can be a false savings. Modern incubators are quite complex machines, electronically speaking, so there's a good deal that can go wrong with the sensitive control and monitoring systems on these machines. Although these defects may be subtle to the point of invisibility to the casual observer, they can be sufficient to upset the running of the machine in ways vital enough to affect its reliable performance. Good incubation is all about constancy of settings, so failure in this respect makes a machine worse than useless.

Parts of an egg

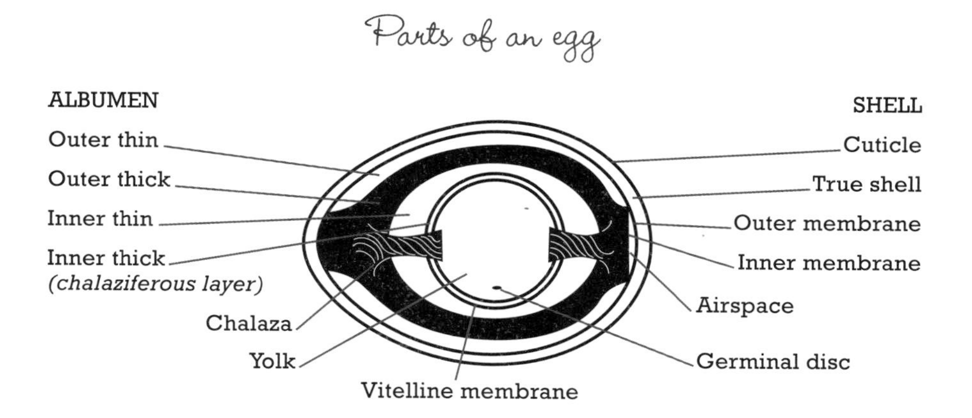

TIP 359: *Take plenty of independent advice when buying*

The choice of incubators available these days makes deciding which to buy a potentially daunting and even confusing business. For these reasons, I'd advise talking to as many experienced breeders as you can before making your choice. If you can find people who are actually using the sort of machine you're thinking about, so much the better. What you need is independent, warts-and-all feedback about how a machine works in the real world. If you can't find a user among your friends or fellow poultry club members, then the next best thing is to join a poultry forum. The Backyard Chickens Forum is one of the best and, among its thousands of members, there are bound to be incubator users with useful stories to tell.

TIP 360: *Buying a starter kit makes a lot of sense*

Some of the better incubator manufacturers are now offering "starter kits" aimed specifically at the novice user. These are carefully put together to provide all the correctly specified equipment needed to get a beginner up and running with the incubation process. Typically they will include a small incubator, egg cleaning fluid, a candling device, a heat source for the brooder, plus chick feeder and waterer units. Thankfully, these days, customers can expect a decent level of technical guidance, plus plenty of back-up support from the manufacturer/supplier via helplines and informative websites, if required.

HATCHING-EGG QUALITY IS EVERYTHING

TIP 361: *Buy your hatching eggs with care*

It may sound obvious, but the quality of the hatching eggs you load into your shiny new incubator will have a direct bearing on the wellbeing of the fluffy little cheepers that emerge from the broken shells after the three-week incubation process. As with that old maxim—garbage in, garbage out—substandard hatching eggs will more than likely result in low hatch rates and unhealthy, poor-quality chicks. So, rather than wasting your time and money, put some effort into sourcing your hatching eggs to help ensure that the results you get meet your expectations. Buy from recommended and respected breeders and you'll get off to the best possible start.

TIP 362: *Watch the condition of home-produced eggs*

If you're using your own home-produced eggs for incubation, then take care at the selection process and pick the best examples you possibly can. For starters, always use the cleanest eggs and avoid those that are badly soiled. Shells that are contaminated with fecal matter will simply promote the development of potentially harmful bacteria inside the incubator. Ideally eggs shouldn't be washed; it removes the protective cuticle from the shell, leaving them more open to contamination and potential hatchability problems. You can buy specialized sanitizing products that must be used exactly as recommended. Never submerge eggs in water to clean them, and always allow shells time to air-dry thoroughly before setting. Never store eggs awaiting incubation in the fridge. Keep them in egg trays, at room temperature and out of direct sunlight.

TIP 363: Take the time to visit the supplier

In many respects, the internet age in which we now live offers unprecedented levels of convenience. However, as far as poultry keeping is concerned, it can be a bit of a double-edged sword. The fact that it discourages people from getting out and actually meeting others is a real disadvantage when it comes to buying hatching eggs or hens. Talking face to face, assessing the environment, looking at parent stock, and generally getting a feel for the levels of husbandry being given are all key pointers from which to help judge the quality of potential purchases. However, buying from an online auction, or a breeder's website, cuts you off completely from these valuable inputs, which is never a good thing.

TIP 364: "Candle" your eggs at the right time

Candling is a simple technique that involves nothing more than shining a bright light through the egg to highlight any development that may be occurring inside. At the first candling session—which shouldn't take place until the eggs have been incubating for ten days—it should be possible to see a spider-like shadow in the center of the egg, assuming all is well. A dark blob in the middle is the embryo, and the "spider's legs" are the developing blood vessels. If there's nothing at all to be seen after 12 days of incubation, then the embryo hasn't developed for some reason, or the egg was infertile in the first place. These eggs (known as "clears") should be removed from the incubator.

TIP 365: *DIY egg-candlers can work just as well*

The relative simplicity of the egg-candling process means that you don't have to spend money buying an actual candling unit. It's perfectly possible to construct your own unit using nothing more complicated than a cardboard tube and a flashlight. The important thing is that the tube diameter is a little less than that of the egg being checked, and that the flashlight is both bright and well attached to one end of the tube with decent sticky tape. Things are made a lot easier if you can find a flashlight with a lamp diameter that more or less matches that of the cardboard tube. Candling itself (best done in a darkened room) should take only a few seconds per egg.

TIP 366: *Fertility doesn't guarantee a successful hatch*

Regrettably, not all eggs you set in an incubator will produce happy, bouncing chicks. Because the incubation process is such a balancing act, in terms of temperature, humidity, and egg turning, there's plenty that can go wrong to affect the end result. It doesn't take too much of a deviation from the required parameters to trigger detrimental consequences. Dramatic variations inside the incubator may halt embryo development altogether, while failure to get things right as hatching time approaches may result in the chick failing to get out and dying in there; a sad and unfortunate outcome known, appropriately enough, as "dead in shell."

TIP 367: *Always mark your hatching eggs*

Good hatching is all about being organized, and a big part of this involves keeping effective tabs on your eggs. Not only is it important to know the breed from which they came, but you must also be aware of the date they were laid. Write this information carefully on the shell using a soft pencil. It's important that this is all you use because, as the shell is porous, there's a risk that any ink-based marker will be drawn through to potentially contaminate and/or damage the contents within. As plenty of eggs have similar-colored shells, it's all too easy to get muddled up about which was laid when, and by what breed of bird. So clear, straightforward shell marking makes a lot of sense.

TIP 368: *Be sure about the provenance of your eggs*

The sad reality nowadays is that there are increasing numbers of people jumping on the chicken-keeping bandwagon, in the hope of turning a quick profit. This certainly extends to the sale of hatching eggs, which are either known to be infertile, or are not the breed being advertised. Dishonest practices like this are getting more common on the internet, with the inevitable time delay between the purchase and the discovery that, one way or another, things aren't as they should be, playing right into the hands of the sellers. If the eggs prove infertile, the supplier simply blames the shaking the eggs got in transit, and it's all but impossible to prove otherwise. What's more, most keepers who hatch birds that are the wrong breed or poor examples aren't inclined to complain so long after the initial purchase. The fraudulent sellers win either way, so be warned.

TIP 369: *Egg shape and size are important too*

One thing that many newcomers to hen keeping fail to appreciate is the variety of egg shapes and sizes that can be produced by backyard chickens. Spoiled by the quality-control efforts of modern retailers, the buying public never gets to see the abnormalities that occur. However, hen keepers soon become aware that you get big ones, small ones, elongated ones and, sometimes, almost perfectly round ones. While these may be a source of curiosity for many, those serious about incubation should avoid using anything other than traditionally shaped eggs. The chicken egg's shape has evolved over millions of years to be most suited to the chick that has to develop and grow inside it. So incubating abnormally shaped eggs simply reduces the chances of a successful hatch. Likewise, eggs that are unusually large or small are best avoided, too.

TIP 370: *Shell quality is another useful indicator*

The shell of a good hatching egg must be naturally clean, free from cracks, holes, and wrinkles. Any defects of this sort should be enough to put you off using that particular egg for hatching. There's simply no point in knowingly using eggs that are outwardly defective as it's likely that there will be problems inside, too. Poor quality shells are typically caused by disease, especially in hens at the end of their laying year or that are getting old. Eggs with wrinkled shells are indicative of overweight birds, and those with faint lines running along the length of the shell suggest disease.

YOUR FIRST HATCH

TIP 371: *Take things slowly and methodically*

You—and probably the rest of your family—are bound to be excited as hatching time approaches. The prospect of each egg breaking apart to reveal a fluffy little ball of joy is enough to warm the heart of even the most cynical old breeder. However, it's very important not to let your enthusiasm get the better of you. Egg incubation is an exacting process that, if it's to be successful, requires a reasonably precise set of requirements to be met. If you have a half-decent incubator that's in good working order, then this should manage the process well, assuming you've positioned it appropriately and followed the operating instructions to the letter. Incubation is a process that can't be rushed and that requires a methodical, considered approach.

TIP 372: *Check everything carefully before you get going*

You can have the best hatching eggs in the world, but if the incubator isn't working properly, or you've made a basic error with its setup, then a good hatch is unlikely. This is why it's so important to check your equipment carefully before you start. Three weeks is a long time to wait, only to discover that the machine wasn't maintaining the correct temperature, or that the humidity was wrong. Always make sure you've run the unit for a couple of days just before you begin, checking that all systems are functioning as they should be. Also, the incubator needs to be clean. Even a brand-new machine will benefit from a disinfectant wipe before you set the eggs.

TIP 373: *Make sure you have brooding facilities ready*

With all the excitement associated with your first go at incubation, it's all too easy to overlook the need for a brooder: a secure and warm environment into which the chicks are transferred once they've hatched and are thoroughly dry. Some of the more expensive incubators can double as brooders once hatching is complete, but most at the DIY end of the scale don't offer this facility. So it's likely that you'll need to make your own arrangements. You can buy custom-made units (an expensive option) or make your own. The DIY approach tends to suit most who are new to this aspect of the hobby, primarily because it's much more affordable. If you're planning to breed a lot of chicks over several years, building a brooder out of wood makes sense as it'll last. However, if your hatching is going to be more sporadic, then an appropriately sized cardboard box and suitable heat source is probably as good as anything.

TIP 374: *Be prepared for the odd disappointment*

The nature of egg incubation, and the large number of variable factors, mean that successful hatching is by no means guaranteed. Everything from the quality of the eggs you're using to the setup and sophistication of your incubator has the potential to upset the hatching apple cart, if you'll forgive the mixing of my rurally themed metaphors. Also, if you have children who are excited at the prospect of their first chicks, it's well worth taking the time to explain to them that things don't always go well and that sad things sometimes happen. It can be very distressing for youngsters when chicks don't hatch, or die soon after struggling to hatch out.

TIP 375: *Even fertile eggs offer no guarantee of chicks*

It's a regrettable consequence of the unpredictability of incubation that eggs that were confirmed as fertile at the tenth day candling can simply stop developing for no apparent reason. Development that may have looked so encouraging at the start can cease for any number of subtle reasons: the incubator may be delivering an erratic temperature, the humidity level inside the machine may be wrong, or the egg itself may be diseased or contaminated, having been sourced from poor quality stock. So, just because you spot the first signs of embryo development at the first candling session, don't assume that you're home, safe and sound.

TIP 376: *Keep young meddlers under control*

Incubators can prove irresistible as far as youngsters are concerned. So, if you have children in the family, it's a good idea to make sure that they are aware of the need to keep fiddling to an absolute minimum. Apart from cursory checks of temperature and humidity levels, and keeping an eye out for any warning lights that may be illuminated, incubators need to be left alone to work their magic. Remember that every time the lid is opened the carefully controlled environment inside is upset. Keep doing this to candle the eggs more than you need to, or to get a better look at what's happening inside, and you'll dramatically affect the temperature and humidity settings; possibly to a degree that's beyond the unit's recovery.

TIP 377: *Get your incubator settings correct*

While most incubators these days are supplied with a temperature readout, so that it's simple to be sure that the required 100°F (37.8°C) is being maintained, assessing the humidity level can be a different matter. Unless your machine features a dedicated level display for this too, you'll need to buy an appropriate humidity gauge to be sure about what's going on inside. At the start of the incubation period, humidity needs to be somewhere between 40 and 50%, and achieving this may require the addition of water inside the machine; much depends on the season. The instruction manual provided with the incubator should offer practical, model-specific guidance about this, as well as details about how to increase humidity as hatching time approaches.

TIP 378: *Give any stragglers a bit of time to hatch*

Although the standard incubation time for a chicken is twenty-one days, it doesn't always follow that the chicks will hatch right on time. It's very rare to have early hatchers, but lateness is relatively common. One or two days isn't usually an issue, but any longer than this should be a cause for concern. Some keepers opt to "help" the stragglers out of their shells in a process known as "assisted hatching." But others are very much against this sort of intervention. They feel that if, for whatever reason, the chick isn't strong enough to get out of the shell on it's own and on time, then there's a problem. Helping under these circumstances is something of a pointless exercise, as all you'll be doing is bringing a "defective" bird into the world that, under normal circumstances, wouldn't have survived. It's an emotional issue that fires plenty of debate.

TIP 379: *Allow hatched chicks time to dry properly*

You'll notice that when chicks first emerge from the shell they're wet, sticky, and tired. It's typically an energy-sapping struggle for a chick to chip its way out, and it's very important that those that manage it are given sufficient time to dry out and regain their strength. This will typically take 24 hours, and the chicks will survive this period by consuming the remains of the yolk sac. Never consider removing newly hatched birds from the safety and warmth of the incubator until they appear completely dry, with fluffy, puffed-up feathering (down). Also avoid unnecessary opening of the incubator lid while the chicks are still damp (especially if ambient temperatures are low), as they're prone to chilling at this delicate stage.

TIP 380: *Make the move to the brooder with care*

Very careful temperature management is required when you move the chicks from the incubator to the brooder unit. Not everyone is able to have the incubator and the brooder in the same room, or even the same building. If this is the case with you, then think carefully about how you transfer the delicate youngsters between the two. Some experienced breeders go to great lengths to ensure their chicks experience minimal temperature change during this important transition, and there's anecdotal evidence to suggest that birds that get chilled never do as well as those that don't. I've heard cases of breeders pre-warming wood shavings in the oven, putting it in a plastic bucket, carefully adding the chicks and then covering the bucket with a thick, warmed towel. They also take great care to make sure that the brooder is at the same temperature as the incubator the chicks have just left.

CARING FOR CHICKS

TIP 381: Don't let young chicks get dehydrated

Once dry and successfully moved into an appropriately heated brooder, the important rearing process can begin. Fundamental to this is that you start the youngsters eating and—especially—drinking as soon as possible in their new surroundings. It's the intake of fluids that's the more important of the two at this early stage. Dehydration is a very real risk, and the damaging effects of it will quickly take hold if chicks aren't given a supply of fresh, clean water at the earliest possible opportunity. Also remember to make sure that waterers are shallow, to avoid the risk of accidental drownings. Ideally use a specially made chick waterer; if you haven't got one of these a shallow bowl or dish will do. If you're at all worried about the depth of water, then add a layer of clean stones or marbles to make it shallower and impossible to become submerged in.

TIP 382: Take initiative if chicks are reluctant to eat or drink

One of the few practical disadvantages associated with using an incubator, rather than a broody hen, is that the resultant chicks lack the valuable teaching influence of the mother hen. Not only does her input give the youngsters a sort of streetwise savvy that encourages them to get out and about more quickly than they might otherwise do, but she also leads by example when it comes to feeding and drinking. If yours are reluctant, dip their beaks in the water to show them what and where it is, then spread some chick starters on sheets of paper and tap at it with your fingertip, to mimic the action of a pecking hen. If you still have no luck, try giving them finely chopped hard-boiled egg, which is both highly nutritious and an attractive color to them.

TIP 383: The brooding period poses many challenges

In many ways, the brooding period of a young chicken's life is as important as the incubation itself. Many chicks are lost during this influential time due to a number of relatively common reasons: insufficient heat, over crowding, too much heat causing dehydration, and dirty or damp conditions in a brooder unit itself. What you must appreciate is that chicks are very vulnerable at this stage. Disease can develop and spread very quickly, especially respiratory problems typically caused by poor conditions inside the brooder. One of the main killers is coccidiosis, which can easily wipe out a brooder-full of chicks very quickly if it isn't spotted and dealt with at an early stage. The classic signs to watch for are listlessness in the chicks and blood in their droppings.

TIP 384: Bear in mind the space you have available

It isn't disease that poses the only threat to your vulnerable young chicks; problems can also be caused by much more straightforward, physical factors, too. Overcrowding can be a deadly issue. In their enthusiasm to eat, drink, and generally be merry, it's quite possible for the less mobile chicks in a group of youngsters to become overwhelmed by the rest. In the worst cases, this can result in smothering and death. For this reason, it's very important that you don't confine too many chicks in a brooder that's too small, and that the unit's shape doesn't allow vulnerable birds to get trapped in corners.

TIP 385: *Leg problems can be an issue with young chicks*

The delicate nature of young chicks and, in particular, the softness of their bones, means that they can be prone to slipping and dislocations. Smooth incubator trays or a brooder floor that are too smooth can be a common cause of an unpleasant condition called "splayed" or "spraddled" leg. The chicks literally slip and do the splits, which leaves them with legs fixed at odd and unnatural angles. Putting this right means gently working the legs back into the correct position, then using a small figure-of-eight of string or rubber band to hold them together until the muscles are sufficiently strengthened to retain them in the correct position. With care, most chicks can be nursed through this condition. Putting corrugated cardboard on the incubator floor will give the youngsters better grip.

TIP 386: *Make sure you feed what the chicks need*

It's important that chicks start eating from the moment they get into the brooder, as their own reserves of energy will be running low by this stage. You must always feed a good-quality chick starter that's fresh and has been supplied by a reputable specialist. This will ensure the birds get all the nutrients and vitamins needed to get the best start in life. The physical size of the chick starter granules makes it easy for the young birds to pick up, swallow, and digest. They should be kept on this for about eight weeks, after which they can be switched over gradually to chick grower. Continue with this through the rearing period until the birds mature and reach point-of-lay, at which time you can start feeding them a layer ration.

TIP 387: *Be on your guard for breathing issues*

Respiratory trouble is unfortunately common among young chicks hatched by inexperienced breeders. Thankfully, spotting the danger signs is quite easy, whatever your level of experience. If you notice chicks standing still, with their beaks opening and closing and apparently gasping for breath, then this is a good indicator that all's not well (assuming that the brooder isn't too hot). While there are many possible causes, the most common culprit is damp and/or dirty bedding. Chicks sleep on the floor and breathe in the ammonia produced by bedding material that's contaminated with droppings. Damage to their sensitive young lungs occurs all too easily. Damp litter can also promote the production of dangerously infectious spores. Treatment for the consequential respiratory problems is available from your vet, and is typically easy to administer via the bird's drinking water.

TIP 388: *Temperature remains key during the early days*

Getting the ambient temperature right for your growing chicks is another vital factor; mistakes at this stage can have serious consequences. Birds that are kept too hot will experience detrimental affects to growth rate and overall health, and they will be more susceptible to disease as a result. Dehydration will be an accompanying and dangerous possibility that will curb appetite and development. Equally, low temperatures can be just as serious; perhaps even more so. Young chicks are very intolerant of the cold, are easily chilled and will die quickly if there's insufficient heat provided. What's more, a chilly environment will cause huddling as the chicks attempt to keep warm, and those in the middle of the group can be suffocated in the crush.

TIP 389: *Watch humidity levels to avoid sticky situations*

Not all chicks hatch as well as each other. Sometimes you'll find with a batch that the chicks emerge looking much messier, with egg material and shell stuck all over them. The most likely causes of this appropriately named "sticky chick" condition, are excessive humidity inside the incubator and/or a marginally low temperature during incubation. Thankfully, this isn't an immediately serious problem, but there can be long-term implications. Sticky chicks tend to be large, "blobby" creatures, often lacking balance and are typically slower than normal to get going. The condition can be avoided by providing the correct incubator environment. Hatches that emerge "sticky" may benefit from being kept a degree or two warmer in the brooder for the first few days.

TIP 390: *Resist the temptation to hatch too many chicks*

Most hatches will produce a 50:50 split between male and female birds, which, in evolutionary terms, is all well and good. At a domestic level, however, it can be far from ideal. Male birds can spell trouble in many ways for the hobby keeper, and the number likely to be produced should be a significant factor to bear in mind for anyone contemplating home-hatching. Once grown up, male chickens tend to be noisy (often at inappropriate times), which can pose all sorts of problems for urban keepers. This aspect undoubtedly contributes to the generally low value of male birds on the open market, making them relatively hard to dispose of through the normal channels. So always think further down the line when setting your eggs: it's important to have a strategy in place for dealing with unwanted male birds.

EFFECTIVE BROODING

TIP 391: *You can improvise with brooders to save money*

There are plenty of options when it comes to creating your own brooder unit, from the very basic to the high end and complex. In most cases, though, the choice comes down to something of a compromise, being determined essentially by the number of birds involved and the budget available. At the bottom of the sophistication scale is the simple cardboard box. This can be perfectly adequate if you're only dealing with a handful of chicks, assuming it's sturdy, secure, and doesn't present a fire risk. Alternatively, you can make good use of an old wooden drawer, although the usual shallowness will mean that a hardware cloth cover will be necessary to prevent the chicks from escaping.

TIP 392: *Adopt the "plastic sheet" approach to brooder building*

A practical alternative to the box-shaped brooder approach is to use a 6 x 3ft (1.8 x 1m) sheet of strong, flexible plastic, which you bend round and fasten, to create a circular enclosure. This is easy to use and simple to keep clean. Also, there are no corners to worry about, so chick-smothering won't be a factor. There is a risk of young chicks getting trapped in the corners of conventionally shaped brooders (especially if the unit is too hot or overcrowded), with suffocation sometimes being the unfortunate result. For this reason, it makes a lot of sense to adapt square-shaped brooder boxes so that the corners are rounded off. This is easy and cheap to do; curved sections of stiff card stuck across the corners is perfectly adequate.

TIP 393: Effective brooder heating is essential

Chicks must be transferred into a brooder that's heated to the same temperature as the incubator they've come from. Basically, there are two common heating options available: a heat lamp (either infrared or white) or something called an "electric hen." Whichever you choose, setting the initial temperature is something that must be done carefully, using an accurate thermometer. Thereafter, though, you can use the behavior of the chicks to gauge how suitable the heat level is. If they're all away from the heat source, around the perimeter of the enclosure, then it's probably too hot. Alternatively, if they're all huddled together, directly under the heat source, then it's too cold. Ideally, the birds should be active and evenly spread around the floor of the brooder unit.

TIP 394: Electricity is a convenient brooder power source

Electricity is the most obvious power source for heating brooders, and it's typically used to run light bulbs of one sort or another. The two most popular options available offer variations in purchase price, power output (100–250W), and overall running costs. Straightforward, white-light or infrared heater bulbs tend to be the most widely used types. Both produce light and heat and, ideally, should be mounted within a specially made housing that incorporates an effective safety guard to prevent accidental contact between the bulb and anything else that might burn. Usually these sorts of bulb are suspended on a chain so that their height above the birds—and thus the heat being delivered—is easy and quick to adjust.

TIP 395: Brooder light control may be beneficial

Light-emitting bulbs aren't the only option; some breeders prefer to split the light and heat sources so that they can be controlled independently. There's a belief that giving young birds a day/night-type light cycle is important, but this requires separate sources as the heat must be provided constantly. One option for achieving this is to buy a "dull emitter." Effectively, these are light bulbs that, instead of a clear/pearl glass envelope, are coated with a ceramic layer; the filament inside generates heat but no light. They offer a much longer service life than a conventional bulb, are correspondingly more expensive, and have power outputs of up to 250W. Although they might be difficult to find in the US, you can find them for sale on Ebay.

TIP 396: Perhaps try an electric mother hen in your brooder

The oddly named "electric hen" represents an interesting alternative to a dull emitter, as a non-light producing heat source. It makes use of a large, heated pad that stands in the brooder on adjustable legs. Its height above the floor can be adjusted to suit the size of the growing birds. The idea is that the chicks can move under the pad whenever they feel the need for some warmth, much as they would do with a broody hen. Some breeders prefer to use this method, believing that it offers a more "natural" experience for the chicks. However, these devices are perhaps best suited to larger brooder units, as they take up floor space. It's important the chicks always have enough area away from the unit in which to feed, drink, and move. There will, of course, still be a need for a separate light source.

TIP 397: *Attention to detail counts when brooding*

The way young birds are reared has a significant affect on both their general health and their overall quality once they are fully grown; it's a vital stage of life that needs to be carefully and correctly managed. At the most basic level, it's essential that chicks are kept clean and dry at all times if problems are to be avoided. Over crowding is something that must be avoided too and, while none of this is rocket science, successful results will require attention to detail. One of the fundamentals to ensure is that the chicks never get damp, as this is a serious promoter of disease. Keep an eye on the waterers in the brooder, making sure that the birds' inevitable clumsiness doesn't cause damp patches to develop in the litter.

TIP 398: *Keep things fresh, but draft-free*

One of the secrets of good brooding (to help guarantee healthy development and minimize the risk of respiratory problems) is to ensure young birds are reared in an environment with plenty of fresh air, but no drafts. This calls for careful placement of the brooder in a calm, dry, and well-ventilated environment, but out of direct sunlight. It's also wise to keep adult birds completely out of the way at this stage too, until the chicks have had time to build up some natural immunity of their own. Remember that older birds can be carriers of all sorts of conditions, even though they may not be showing any outward signs of illness. Any health "challenge" faced by vulnerable chicks during the first few weeks of life can prove fatal, so it's never worth taking chances.

TIP 399: *Take great care with your introductions*

As the chicks you've hatched and reared grow steadily larger, you may be tempted to simply introduce them to your existing flock. However, think carefully before acting because this can frequently be a cause of problems. The rules to observe include trying to ensure that you put compatible breeds together, in terms of age, size, and type. Mixing different sizes of bird will often cause problems, including bullying and feather picking (both of which can prove fatal). Also remember that obvious physical, breed-related differences are likely to provoke trouble too. For example, crested breeds don't mix well with noncrested types.

TIP 400: *Make your move after dark*

Introducing "new" birds from the rearing pens to the existing flock can be a problematic business. Anything that upsets the existing pecking order within an established group of birds is likely to put a good few beaks out of joint. So adding new birds requires care and sensitivity. The best advice is to add the newcomers after dark, once the rest of the flock is roosting quietly in their coop. Carefully place the new birds inside the coop, disturbing the residents as little as possible. They should all then emerge in the morning on good terms, although you'll need to remain on hand afterwards to keep a wary eye on proceedings.

TACKLING COMMON PROBLEMS

Chickens are a prey species, which means that they do an amazing job in disguising their weaknesses, appearing outwardly fit and healthy. Great for fooling potential predators looking to pick off stragglers from the flock, it's not such good news in the domestic setting. This ability to mask illness and infection makes it more difficult for keepers—especially those new to the hobby—to recognize when things are going wrong. But problems do occur and it's important that you take action quickly to start to remedy the situation.

EXTERNAL PARASITES

TIP 401: *Never lower your guard against red mites*

Red mites are probably the most common and potentially disruptive ectoparasite affecting domestic poultry today. Whether it's the proliferation of newcomers to the hobby; the inability of domestic chicken keepers to provide effective control measures, or the generally warmer winters we seem to get these days offering a kinder environment—or a combination of all three—these pests are most definitely on the march. With all this in mind, the importance of doing everything within your power to keep these worrisome creatures under control is paramount. They are superbly adapted for taking advantage of lapses in your defenses, so don't give them a chance to do so.

TIP 402: *If you can spot them, you've got a real problem*

The effects of red mites, especially in the early stages, can be tricky to spot if you're not entirely clear about what you're looking for. The creature itself is never particularly easy to see; it's small and tends to be most active at night. Mites you find crawling about on your birds during the day are likely to be another parasite, the Northern fowl mite. So, unless you're prepared to get outside after dark with a good flashlight and a pair of gloves, it's unlikely that you'll ever see red mites in action. The one exception to this is actually the worst-case scenario: when the number of these little pests reaches infestation levels. Under these conditions, the need for food will force the critters out of their usual hiding places on daylight food raids, and they'll often be visible to the naked eye. Not a welcome sight.

TIP 403: Be ever-watchful for signs of trouble

By choice, red mites operate at night, emerging from within the structure of the hen coop, crawling up on to the roosting birds, and settling down to feed contentedly on their blood. This, as you will appreciate, is both irritating and depressing for the victims. However, hens that are young and in good general health will be able to withstand a degree of this kind of parasitic plundering. It's only when mite numbers reach infestation levels that serious problems can arise, even for the fittest of birds. Hens that are forced to endure a serious, blood sucking onslaught night after night will go downhill fairly quickly and, as a consequence, egg numbers will fall and they'll become anemic. In severe cases, birds will be reluctant to re-enter the hen coop in the evening and, if not dealt with, the downward trend in health will lead to death.

TIP 404: In the nastiest cases, you can actually smell trouble

In coops where there's a heavy red mites infestation, it's actually possible to pick out a characteristic but hard-to-describe stale, "dirty" smell. Expert keepers who have experienced it have said that it's not something you'll ever forget afterwards. Of course, if you ever find that things are this bad, then drastic and immediate measures will be called for. You'll need to move the birds out of their infested coop, treat them individually with a good antimite product, and find them some new lodgings. In practical terms, the best place for the old coop is probably the bonfire.

TIP 405: *Mites can be an issue for broody hens, too*

If you're hoping to hatch chicks as Mother Nature intended, using a broody hen, it's important not to forget that mites can be a serious issue for birds in this situation too. If you're expecting your hen to sit contentedly on her clutch of eggs for three weeks, you must do your bit to ensure that the environment is comfortable. Anything that's going to cause irritation or stress is likely to make her desert the eggs, and that'll be that. So it's essential that you take appropriate measures against lice and mites.
The broody box she's going to sit in will need a thorough cleaning and disinfection (especially if it's been used before). It should then be dusted with antilouse powder before the bedding goes in, and the hen will benefit from a decent sprinkling too. The presence of parasites can quickly weaken chicks, lowering their resistance, and increasing the risk of other disease problems.

TIP 406: *Lice require a close inspection*

Lice are impossible to detect from any distance, and will only be apparent after a thorough inspection of the bird while it's "in hand." These small parasites move about on the skin under the feathers and can be quite tricky to track down. Typically they will congregate around the bird's vent, and their white eggs (like nits in hair) can be found stuck in clusters on to the base of feather shafts. There are several species that can affect chickens and, believe it or not, different types may be discovered at various locations around the body (on the abdomen, around the vent, at the base of the tail, on the wings, and around the neck).

TIP 407: *Never underestimate the tiniest pests*

Mites are minute, parasitic insects with eight legs, which suck the blood or eat the body tissue of their host. In the case of chickens, the birds suffer from the irritation mites cause, pain and/or anemia and, in cases of severe infestation, may even die. Red, scaly leg, depluming, and Northern fowl are all types of mite that could detrimentally affect your chickens. There are numerous powders for treating the birds and/or the coop and its fittings. These should always be used as per the manufacturer's recommendations, and treatments repeated to kill mites that hatch following the initial application. Unfortunately, mites become resistant very quickly to many of the chemicals commonly used; if a treatment fails, it's worth seeking veterinary help in sourcing prescription-only alternatives.

TIP 408: *Keepers of crested breeds, watch out*

There's some evidence to suggest that crested breeds seem to be more vulnerable to infestation with Northern fowl mite than other types of chickens, so keepers of birds such as the Poland need to be extra wary, carrying out careful examinations, especially if the birds have been taken to compete at a poultry show. These pests look similar to red mites; the two parasites are closely related, with both being small and reddish-brown in color. However, Northern fowl mite live on their host permanently and have the potential to cause serious problems and even death. Hens suffering with a heavy burden will be lethargic, may display pale wattles and combs and also "soiled" feathers; the mites' feces contains blood that stains the feathers, which is most noticeable on light-colored birds.

TIP 409: *Protect your hens from over-amorous males*

Those of you who keep male and female chickens together will find that the roosters often adopt just one hen as their favorite. Typically this is bad news for the female as, during the breeding season, it means that she'll be pestered by the male, who will delight in "treading" her many times every day. Treading is the term given to the mating act, when the male mounts the female and uses his beak and claws to hold on. While this is healthy most of the time, painful problems can arise for a hen that's being repeatedly picked on. Birds that live in a restricted area can be more prone to this sort of problem: the hens simply don't have enough room to get away from amorous males.

TIP 410: *Use a poultry saddle to prevent serious injury*

If you notice that one of your hens is being "abused" by a randy male bird, and there's damage starting to be done by his claws and beak, you can't simply ignore it. If it's not possible to separate the birds, then another option is to treat any wounds and then carefully fit the hen with a poultry saddle. This is a simple canvas cover that fits over the hen's back, and is held in place by straps that loop around the base of the wings. It effectively protects the back and flanks from claw damage, but it's very important to realize that birds fitted with saddles need regular checking for—and treatment against—lice and mites beneath the cover. This should be done on a weekly basis because the area under the saddle offers perfect sanctuary for unwanted creepy-crawlies.

INTERNAL PARASITES

TIP 411: *Prevent problems: instigate a worming schedule*

Prey animals, creatures vulnerable to attack from predators, make strenuous efforts to appear fit and active at all times; the last thing any of them want is to look weak and immobile as this will mark them out as an easy meal. With chickens in a domestic environment, though, this primitive ability can work against the keeper, who needs to be able to recognize the signs of ill health among his or her birds at the earliest opportunity. Things can be made doubly hard when the problem stems from a condition within the body, such as a parasitic worm infestation. Outward signs of the internal struggle that's raging inside will remain minimal, which is why many keepers are preemptive with their worming program.

TIP 412: *Understand the roundworm risk*

There's quite a range of worms that will happily take up residence inside all species of poultry, but the largest of them is the roundworm. These are stout, thick, white worms, and the females can measure up to 4¾in (12cm) long. Infection is caused by the consumption of eggs via fecally-contaminated food, soil, or grass. It takes six weeks for the swallowed egg to hatch and then develop into an egg-producing adult in young birds; this takes eight weeks in adults. The worm eggs in droppings on soil may be ingested and carried by earthworms, so hens that free-range may be
at greater risk.

TIP 413: *A fixed run may create a worm breeding ground*

The number of worm eggs present in the soil of a run can rise quickly due to the relatively short time between the infection of a bird and the commencement of egg production in its feces. This can be a particular problem in small runs in a fixed location if the droppings are not collected and removed. The number of infective eggs available to and eaten by the birds will continually rise, and thus the internal burden placed on the birds will steadily increase. The common signs of roundworm presence can include weight loss, diarrhea, huddling, feather fluffing, and even death if a bird is heavily infected. Moderately infected birds may just be "slow," showing slow weight gain, and not being as alert or quick as the other birds.

TIP 414: *Effective roundworm control is important*

Hens with a serious level of roundworm infestation can suffer with a blockage in the intestine, which may be sufficient to lead to death. What's more, badly affected birds can also suffer even when efforts are made to make things better. Worming a hen that's suffering with a heavy roundworm burden can do more harm than good; a large number of worms die off at the same time as the wormer takes effect, and these can actually get knotted together and cause an internal obstruction. Thankfully, though, chickens develop an increasing resistance to roundworm as they get older. The intestinal environment changes with age, becoming less favorable to worm survival.

TIP 415: *Get the conditions right to minimize worm numbers*

Once hens get past the three-months-old mark, the effects of a worm burden should be more easily managed. However, poor environmental conditions remain a vital factor, both in controlling worm numbers overall, and by helping to ensure that this natural level of resistance is achieved. Wet areas around waterers, or swampy ground in the run generally, can increase the numbers of infective eggs surviving in the ground or coop bedding, thus increasing the chances of the birds becoming heavily infected. Soil-floored enclosures will naturally allow access to worm egg-carrying earthworms, and there will be all the more of these if conditions are generally wet.

TIP 416: *A balanced diet is vital for a hen's resistance*

Never forget that a poor, badly balanced diet is likely to provide hens with less resistance to the threat of worms so that, if these internal parasites do gain a hold, it'll take fewer of them to start having a truly detrimental effect. A deficiency in vitamin A is regarded as very significant in this respect. The basic, antiworm ground rules are:

1. Feed a well-balanced ration suitable for the birds eating it.
2. Keep runs and cages clean and prevent the buildup of droppings.
3. Feed from a feeder, not the ground.
4. Move soil-based runs regularly on to fresh ground.
5. Worm "at risk" birds regularly with a recognized, medicated wormer.

TIP 417: *The tapeworm is another potential trouble-maker*

Another type of parasitic worm commonly found in poultry is the tapeworm, although these rarely cause a serious problem. The adult tapeworm buries its head in the lining of the bird's intestine; the tail section of the body (which is segmented) fills with eggs and then falls away into the stream of passing matter, and is then expelled from the body via the droppings. The eggs mature out in the environment, are ingested by an insect or snail and then infect another bird when that creature is, in turn, eaten. And so the cycle of life continues. There's one type of tapeworm that, despite only being 3/64 in (0.5cm) long, can cause weight loss, loss of condition of plumage, weakness, and slow growth even when in low numbers. A heavy bout can lead to bloody diarrhea.

TIP 418: *And the tapeworm presents a variable threat*

Some species of tapeworm are quite likely to cause disease in chickens, and so represent more of a potential danger. There are types of *Raillietina* tapeworm found in beetles and ants, which the birds eat to become infected. These can grow up to 10in (25cm) long, and a heavy burden will cause a noticeable weight loss for the bird. Once again, though, there's an age-related immunity to these worms (chicks over 10 weeks old are more resistant). Prevention of all tapeworm infection centers around separating the birds from beetles, snails, and ants—no easy task! These worms are usually treated with broad-spectrum wormers, including fenbendazole.

TIP 419: Take appropriate action when a problem's identified

The only way to be totally sure about whether or not there's a serious worm problem among your birds is to analyze a droppings sample. This can be performed by your local avian vet, or you can send a sample out to a lab. Then, based on the egg count results from the test, your vet will be able to advise about the most appropriate course of action. If the situation looks bad, then it's important that the measures taken are effective ones that are going to sort out the problem in the shortest possible time. With this in mind, chemical-based treatments tend to represent the most practical way forward.

TIP 420: There's a DIY option for worm egg counting

If you're interested in keeping tabs on the parasitic worm populations in your poultry pens, then it's perfectly possible to get a representative idea about how things stand. You'll need a second-hand microscope (100x magnification), a few test tubes and some glass slides. A dropping sample needs to be mashed then mixed in a strong salt solution. The resultant liquid is then poured into a test tube, which is filled as full as possible. Place a glass cover slip over the top of the test tube, so it's touching the surface of the liquid, and leave it for 20 minutes. This gives anything interesting plenty of time to float to the surface and adhere to the cover slip. Then this can be removed, placed on a glass microscope slide and examined under the microscope. You can find out how to identify worm egg shapes and types on the internet.

RESPIRATORY ISSUES

TIP 421: *Swollen faces can spell trouble*

Respiratory disease often leads to chickens getting swollen heads, especially when the sinuses become blocked, mycoplasma is a common cause for this kind of symptom. "Myco" is an increasingly common occurrence among backyard chickens these days, and can only really be tackled effectively with veterinary help; it's not a condition that'll run its course, then go away. It tends to be most frequently triggered by stress in the birds, so good husbandry—with particular regard to hygiene, ventilation, and stocking densities—is the best way to help maintain resistance, and is vital to the recovery of those being treated.

TIP 422: *Always minimize the stress of transportation*

Moving hens from one location to another—from the breeder to their new home, rather than simply lifting them over a fence at home—is a stressful process for chickens, so can be another trigger for mycoplasma. Ideally, then, the transportation of birds should be kept to a minimum. However, it's inevitable if you're keen on showing. To help ease the ordeal, there are some vitamin tonics on the market that can help if given before a known stressful event, although these aren't a guaranteed preventative. Also, any event where birds will mix is always going to be a potential source of infection, so an effective management procedure—including isolation afterwards—is a good way to help ensure you don't bring unwanted guests back from the show with you. Buying new birds from poultry auctions remains the major source of infection of all sorts.

TIP 423: *Take great care with unknown new birds*

If you've sourced some new hens from a poultry auction or livestock sale, and already have an existing flock at home, it's very important that you implement a two-week quarantine period for the new birds. Not only will this give them time to settle into the new environment in a relaxed and stress-free way but, perhaps more importantly, it'll give you time to assess their general health and pick up and then deal with any signs of disease before introducing them to the main flock. When dealing with the quarantined poultry, it's important that all equipment be kept separate from the main flock, and that the new birds are far away enough to avoid any risk of airborne disease spreading. If you can't manage this, at least prevent beak-to-beak contact between the groups.

TIP 424: *Chickens can catch a "cold"*

Mycoplasma isn't a new disease and, in its time, it's been known by a number of other names, including "bubble eye" and "cold." The organism itself is a type of small bacteria that lacks a cell wall, which makes it a very tricky customer to treat with antibiotics. This is one of the main reasons why it continues to be such a persistent threat. There are seventeen different avian strains, but only four of them affect domestic poultry. The clinical signs of mycoplasma to watch for can vary depending on the strain, but the classical "cold"-like appearance (runny nose, sneezing) is generally *Mycoplasma gallisepticum* whereas lameness and hot joints can be a sign of *Mycoplasma synoviae*.

TIP 425: *Mycoplasma is hard to treat, but not impossible*

Unfortunately, the treatment potential for birds suffering with mycoplasma is both limited and expensive. Just a few antibiotics have been licensed for use with domestic chickens and, to make matters worse, even these don't guarantee eradication of the infection. Generally, the best you can hope for is a reduction of the clinical signs of the infection. This makes avoidance of the disease in the first place all the more important. Your vet should also be concerned about the responsible use of antibiotics, and won't want to continue on blindly with the same type if it's not working. "Carrier" birds can exist showing no outward signs of the disease, which is why outbreaks can suddenly appear following a period of stress within the flock. Vaccination is an option, but it's only effective on birds or chicks that haven't been exposed to the disease; most domestic flocks will have a few birds that are carriers.

TIP 426: *IB is one of the most common respiratory diseases*

Infectious bronchitis (IB) is a disease to be avoided as it can trigger a host of secondary infections that will dramatically increase the threat to your chickens. It's caused by an ever-evolving coronavirus that attacks the upper respiratory tract, and can have an incubation period of as little as 18 hours. The infection is highly contagious, and some birds will be carriers for up to a year while others will die. The virus itself can survive for up to a month in the hen coop. Hens enduring poor ventilation or over crowding will be at greatest risk, and the signs to look for include loss of appetite, lethargy, diarrhea, and coughing/gasping. If you suspect this, immediate and knowledgeable veterinary assistance is essential.

TIP 427: *Coryza: it's highly infectious*

This is an unpleasant respiratory condition that causes severe catarrhal inflammation of the mucous membranes in the upper respiratory tract. It's also known as "roup," and is spread, bird to bird, by respiratory discharge. Coryza is caused by a bacterium and the infection may run either an acute or chronic course. The main signs are sneezing, coughing, and difficulty in breathing. The facial tissues and wattles also become swollen, and there's a foul-smelling discharge from the eyes and nostrils. Food and water intake will be reduced, leading to a loss of weight and a fall in egg output. However, mortality is low in uncomplicated cases. Suspected cases should be referred to your vet for effective diagnosis but, in many respects, treatment isn't advisable. This will only alleviate the symptoms, allowing the "recovered" birds to become carriers that will then pass the infection on to others. Regrettably, the only real solution is the culling of affected birds.

TIP 428: *Newcastle disease is certainly not to be sniffed at*

Newcastle disease is one of thirteen reportable diseases in the US, meaning that suspected outbreaks must be reported to a vet or the local Animal Health office without delay. It's caused by a virus and there are mild, medium, and virulent strains that can affect chickens. Thankfully rare, this highly contagious disease is transmitted, bird to bird, via infected droppings and respiratory discharge. Therefore, it can be spread between farms or backyards by visitors, on equipment, via wild birds, or even through the air. The signs to watch for include depression, laboured breathing, wheezing, and gurgling, loss of appetite, and a drop in egg production of 30–50% or more. A short time after respiratory signs, nervous symptoms appear; these may manifest themselves as poor muscular coordination of the neck and legs, body tremors, and shaking, plus total or partial paralysis of one or both legs.

TIP 429: *Your vet must help with suspected Newcastle disease*

The actual effects of Newcastle disease can be variable. There may or may not be mortality; virulent strains can cause severe losses, with depression and death in 3–5 days without any of the respiratory signs. But infections of the mild form may go more or less unnoticed. Egg production, in terms of numbers, returns to normal after 1–2 months among survivors, but the quality is likely to be reduced thereafter. It's essential to get a vet involved to assist with diagnosis, via a postmortem examination. This will be necessary because the disease presents signs that are similar to other respiratory infections. Unfortunately, there's no cure for Newcastle disease; a course of vaccinations represents the best method of control and, once again, advice regarding this should be sought from a specialist poultry vet.

TIP 430: *Avian influenza in hens is a notifiable disease*

There have been various avian influenza (AI) outbreaks around the world in recent years and the problem has touched the US too. It's a highly contagious, viral disease affecting the respiratory, digestive, and/or nervous system of many species of birds, including chickens. There are two types of the virus: low pathogenic (LPAI) and highly pathogenic (HPAI). Within the LPAI types, there's evidence that certain H5 and H7 strains may mutate and become highly pathogenic. Symptoms include a sudden drop in egg production, an increase in shell-less eggs, loss of appetite, and general moping. There are respiratory signs too, such as gurgling, "rattling," and discharges from the eyes and nostrils. In its mild form, the only indication of the disease may be a drop in egg production and the symptoms of a "cold." Mortality rates vary according to the strain of virus, and range from negligible in adult stock to as high as 90% in young chicks. There's no treatment; prevention is by vaccination and biosecurity measures.

EMERGENCY MEASURES

TIP 431: *Pecking damage can be an effect of over crowding*

Most problems with poultry stem from basic mistakes made by the keeper, and one of the most fundamental is over crowding the birds. Hens without enough space to move freely and do their own thing will quickly become irritated and bored. Then, before you know it, they'll be taking their frustration out on each other, with those at the bottom of the pecking order tending to bear the brunt of any aggression. Pecking injuries, especially to the head, can bleed a lot and turn nasty. The presence of blood usually makes the situation much worse, very quickly. The level of violence rapidly escalates unless the injured bird is isolated for its own safety. Flesh wounds are best treated with veterinary wound powder, and must be allowed to heal fully before the bird is reintroduced into the flock.

TIP 432: *Combs and wattles can take a pounding*

It's not only other birds' beaks that can cause damage to a chicken's comb; they can also snag them while rooting around in thick undergrowth or, more commonly, on the hardware cloth that's used on many hen coops and run extensions. Those keepers who opt for triangular, ark-style units can unwittingly be putting their birds at a greater risk of headgear damage, as headroom in these structures can be limited. Also, keepers who make their own run extensions using traditional chicken wire can run the same risk, too. This will be especially so if the birds are tempted to poke their heads through in an attempt to get at whatever's growing on the other side. The same rules apply to comb and wattle damage as to flesh wounds. Isolate, treat, and allow a full recovery before reintroduction.

TIP 433: *Certainly don't let it all hang out*

A prolapse is probably one of the most visually disturbing events that hen keepers are likely to face. It occurs when the end of the oviduct (the tubular organ through which eggs are carried) gets pushed out through the vent, due to the bird straining to lay an egg. It can often appear that the organ has turned itself inside out, and it's always a shocking sight. It's most likely to occur in pullets as they come into lay, especially if they are somewhat undersized birds. Alternatively, a large, double-yolked egg (triggered by a rapid increase in day length) can simply be too big to lay, with the same result. Regrettably, treatment is difficult, especially for the novice. Gently pushing the exposed oviduct back in through the bird's vent is not for the fainthearted, and often results in infection or reoccurrence in the future. Beginners should certainly seek veterinary advice.

TIP 434: *Address bumblefoot early*

When things go wrong with your hens, it's essential to deal with them effectively to minimize any suffering. The presence of abscesses can represent a perfect case in point. These can occur on the feet (a condition known as bumblefoot) or on the breast, where rough roosts have caused an injury that's become infected. If ignored, a bumblefoot infection can move up the leg and into the hock joint to cause even more pain and discomfort. The problem is that it's easy to miss this kind of injury if you're not handling your birds regularly. Bumblefoot may be helped by carefully removing the scab, squeezing out any pus from the cavity, then washing with an antiseptic or saline solution. After this, the bird should be put in a coop on clean, soft litter. The treatment may have to be repeated. If the infection spreads, seek veterinary help.

TIP 435: *Be on your guard against the risk of salmonella*

There are over 2,000 types of *salmonellae* bacteria that can infect most animals, including chickens and humans. Hens can carry some types without showing any ill effects, and can pass them on through direct contact or via contaminated eggs or meat products. The result can be serious infection in humans. Thankfully, most *salmonella* bacteria is killed by cooking; temperatures of 212°F (100°C) for three minutes or so are usually enough. However, it's essential that this temperature is achieved uniformly throughout the meat to ensure complete safety.

TIP 436: *A chicken in the hand...*

Another way to induce undesirable stress among your birds, and to risk unnecessary injury, is to catch and hold them incorrectly. Chickens need to feel safe and relaxed while being held, and this means supporting them so they're balanced. From the front of the bird, slide one hand down the breast bone until your fingers reach the legs. Then, spread your fingers slightly and slide on, so that your two middle fingers go between the legs and the other two extend past on either side. Once the bird's legs reach the joint between your fingers, your hand will be in the right place to support the bird in a balanced way while, at the same time, having a grip on its legs but not crushing them together. If the bird feels unbalanced it will flap in an effort to prevent itself from falling; this is when injuries can occur and stress levels rise. Held properly, a bird should sit quite contentedly on the palm of one hand.

TIP 437: *Home slaughter; it's no easy job*

There are four distinct classifications for the way in which a bird is killed: emergency slaughter, nonemergency culling, private slaughter, and commercial slaughter. If a bird needs to be killed immediately, on welfare grounds, then that counts as an emergency slaughter. If welfare isn't an issue, and the bird is being killed for meat that's to be sold, then you're heading down the commercial slaughter route, where there are many more rules and regulations governing procedures. Be sure to investigate local laws prior to slaughtering. If the bird is diseased, and not considered fit for human consumption, then its slaughter is classified as nonemergency culling. It's important to be clear about the distinctions between the four classifications.

TIP 438: *Stunning and dislocation provide a humane end*

It's essential with any slaughter operation that it be both humane and quick. Ideally, the approach of choice would involve first stunning the bird into a state of unconsciousness, and then killing it immediately using either neck dislocation or bleeding (caused by cutting the throat). In practice, though, the stunning aspect is difficult for the domestic keeper; electrical stunners are expensive and rather scary to use, especially in a home environment. An alternative stunning method can be induced by a blow to the back of the head, but this is difficult to judge; insufficient force will fail to stun and cause extreme distress and suffering. So, most keepers opt for a straightforward neck dislocation. This remains a viable method of slaughter, even without prior stunning, assuming it's performed correctly.

TIP 439: *Neck dislocation is best reserved for emergencies*

When neck dislocation is carried out to perfection, the bird is rendered unconscious instantly, and death follows shortly afterwards. Unfortunately, this isn't the case every time and it's this inconsistency that may result in distressing situations for birds and owners. It may be preferable to use neck dislocation only in an emergency situation, or for the slaughter of a very small number of birds where better methods aren't available. It's very much a matter of good technique; something that's best learned through accomplished demonstration, rather than from a textbook description.

TIP 440: *Neck dislocation must be speedy and decisive*

Once you have the bird isolated, calm, and away from any others you keep, pick it up, take a gentle but firm grip on its legs (and wing tips, if you can, to minimize flapping) with one hand, then turn it over slowly so that the underside of its body is against your thigh. Make the first two fingers of your spare hand into a V-shape and grip on either side of the bird's neck, immediately behind the skull. Position your thumb under the beak and, in so doing, twist the head back slightly. With the bird securely held and gripped in this way, the dislocation must then be caused by a swift pull downwards (stretching the neck) while, at the same time, pressing your knuckles into the bird's vertebrae and pulling the head backwards. It's essential that the whole action is swift, firm, and confident. Decisiveness is vital.

TIP 441: *Don't be shocked by the after-death flapping*

Afterwards, you should expect a certain amount of nervous system-related wing flapping and muscle twitching, which can be very unsettling for the inexperienced; it immediately starts putting doubts in the mind about the effectiveness of the dislocation. This is why it's important to check for positive signs that the bird is actually dead. These include feeling for an obvious gap in the vertebrae (caused by the dislocation), ensuring that it's not still breathing, confirming that the pupil are dilated and that there's no reaction from the eyes when they are touched directly.

TIP 442: *Killing cones: another slaughter option*

Another popular home slaughter method is the killing cone. This is a sheet of metal, folded into a cone shape and suspended from a wall or a board. To use the cone, place a bird in with head pointing down. Then cut through the artery at the base of the jaw. The bird will bleed out and expire with minimal struggles, and you can proceed with the butchering process. The cone immobilizes the bird, which prevents bruising of the meat and also allows the bird to bleed out completely. If you do decide to take this route, it is advisable to seek more information from experienced hen keepers or associations.

POULTRY VICES EXPLAINED

TIP 443: *Chickens can develop bad habits*

Bad habits, such as feather picking, cannibalism, and egg eating, are often started by a single bird within a flock, but are all too soon picked up by the others unless preventative measures are taken. Left unchecked, destructive "vices" like these can escalate into a serious problem in a surprisingly short space of time. However, it's important to realize that, in most cases, it will be management or husbandry failures that trigger the problem in the first place. The exception is a genetic predisposition towards aggression, which can make problems more likely in some breeds or strains than others. Nutritional deficiencies, overcrowding, over-bright lighting, overheating, and a bland environment can all provide the stimulus for feather-pecking and, worse still, cannibalism. Egg eating may start as the result of eggs being laid on the floor because of inadequate nest box provision, or a lack of nest training.

TIP 444: *Deal effectively with persistent egg eating birds*

In the case of an outbreak of egg eating, various remedies have been suggested, such as using eggs filled with mustard or chili paste, or feeding eggs until the birds are sick and tired of them. Things like this are supposed to do the trick but they can't be relied upon. The sad fact is that, in bad cases where the habit has become engrained, the only real cure is to identify the offenders and remove them permanently. This is why it's important to recognize the problem at the earliest stage, before others in the flock start playing copycat. The only other option is to use roll-away nest box floors so that, once laid, the eggs are carried away, well out of pecking range.

TIP 445: *When pecking becomes a problem, act swiftly*

Feather picking is a relatively common problem; chickens have a tendency to lash out at each other with pecks at the best of times, but premeditated feather picking is more serious. At the first sign of this undesirable trait, the offending bird will need to be removed from the group. Casualties should be isolated too, so that their affected areas can be cleaned and treated with an appropriate antibiotic ointment or powder (repeating the treatment as necessary). If caught in time, feather-pecked birds should make a full recovery. Maintaining a comfortable pen environment, increasing the space available as birds grow and feather up, and anticipating the bird's requirements before they become critical will do much to avert cannibalistic problems like this one.

TIP 446: *Floor laying should be discouraged*

Hens that get in the habit of laying their eggs on the floor of their coop are bad news. Not only will this increase the risk of eggs getting damaged or broken as the birds move about inside the coop and jump off the roostes, but it'll also increase the chances of egg eating and shell contamination from droppings. To help avoid the problem, the hens must be encouraged to use the nest boxes properly. They need to be a decent height above the coop floor (but not above the roosting perches), and can be "primed" with golf balls or plastic eggs to help give the hens the right idea. Collecting any floor-laid eggs regularly is another good idea. Also, nest boxes should be positioned out of direct light and fitted with burlap curtains if this isn't possible. Any hens seen laying on the floor should be picked up and put in a nest box.

TIP 447: *Keep an eye on young birds' feet*

Toe picking is an unpleasant form of cannibalism and another nasty vice to which chickens can succumb. A damaged toe, especially if the nail has become detached and there's fresh blood on view, may be all that's needed to start this nasty practice. Likewise, chicks may be attracted to the bright, shiny nails of others in their group, especially if they're in a brooder unit that's too brightly lit and/or lacking adequate floor litter. Treatment for the sufferers involves cleaning the wounds and applying styptic powder to stanch the bleeding, although it's likely that the toe will remain permanently scarred and nail-less. To reduce the likelihood of pecking, pens should be deeply littered with a good-quality bedding material, such as white, softwood shavings, and excessive light levels should be avoided. Generous space allocation is another requirement, while the additional interest provided by suspended cabbage or other green vegetables can be a useful distraction.

TIP 448: *Pick your breeds carefully to avoid trouble*

Hybrid hens tend to be extremely docile birds; they were created that way, as an aggressive character isn't conducive to maximizing laying performance. However, some of the pure breeds can be a lot less predictable. Aggressive tendencies can run in strains, so it's very important when buying that you discuss this aspect with the breeder. This is especially the case if you are going to encourage children to become involved with the birds. Size isn't always the best guide, either. While some of the largest, soft-feathered breeds are as gentle as you could wish for, some of the smallest bantams can be feisty little characters. So pick your breeds with care. Make your choices based on the recommendation of other keepers you know and trust, and try to avoid dramatic size differences among your birds. (See also chapter 3.)

TIP 449: *Always make sure there's room to retreat*

You don't need to have kept chickens for very long to start appreciating how aggressive they can be when the mood takes them. While it's true to say that most chicken-coop clashes are short lived and end pretty harmlessly, the likelihood of this happy outcome can be greatly boosted by ensuring that the birds always have enough space to get out of each other's way. There's nothing like a shortage of space for turning what should be a moment of awkwardness into a potentially serious incident. If the put-upon hen isn't able to turn and put reasonable distance between itself and the aggressor, then there are inevitably going to be problems. The same can apply if food and water outlets are limited. If there are awkward relations within your group of backyard hens, then adding a second or third feeder and waterer could ease tensions significantly. If things don't sort themselves out naturally, then you'll have to think about permanent separation.

TIP 450: *It's sensible to have a "sick bay" standing by*

When things go wrong with poultry, it often happens all of a sudden. For this reason it makes sense to be prepared. In most emergency-type situations, what you're going to need to do first is isolate the affected bird. It's essential that you're able to get an injured or ill bird away from the rest of the flock and into safe and secure isolation. In an ideal world, this would be a separate pen that's out of sight and sound of the main enclosure, but not everyone has room for such luxuries. So, if space is tight for you, then a good-sized second-hand rabbit hutch or dog crate will do perfectly well as a temporary sanctuary. Remember you'll need to clean and disinfect it before use, add a generous layer of fresh bedding material, and provide small containers for food and water. If this "sick bay" has to be in or close to the existing hen enclosure, then facing it away from the other birds can help.

EXHIBITION AND HEN RESCUE

Poultry showing is a specialized little world, populated by dedicated and enthusiastic individuals. The fact that the 100 or more pure breeds continue to survive is testament to the tireless efforts of these people. Without their careful breeding programs and fastidious attention to detail, the range of chicken breeds we have access to simply wouldn't exist. Let's hope that their efforts continue to inspire future generations, so that newcomers to the hobby continue to have a smorgasbord to choose from.

SHOWING INTEREST

TIP 451: Have faith in your abilities

There's no denying that keeping a few hens at home as a source of healthy, fresh eggs is great fun, but there's even more potential for increasing the fun factor if you become actively involved in the poultry showing scene. The good news is that it's neither complicated nor expensive to get started, and doing so will open up a whole new area of interest for you. However, it's worth pointing out right at the start that poultry showing for the inexperienced beginner really is all about the taking part, rather than the winning. With the best will in the world, you can't expect to be successful at your first event unless you're extremely lucky. So, regard your early events as experience-builders. Make friends, ask questions and learn as much as you can from your fellow competitors.

TIP 452: It takes time to reach the highest show standards

Although it's tempting to imagine that your bird is the best there's ever been, and you're going to walk in and wipe the floor with the competition, the reality is usually rather different. The poultry exhibition scene is every bit as competitive as any other type of livestock showing, and many show enthusiasts regard this aspect of the hobby as the most important of all. Breeding birds to match the established standard as closely as possible is the name of the game, as far as they're concerned. These dedicated individuals can spend 20 or 30 years working for nothing but the satisfaction of getting things right. They seek perfection; few ever find it but many get mighty close. However, the level of breeding and selection knowledge they build along the way is something else.

TIP 453: Be realistic about your early show choices

The important thing with poultry showing is not to get disheartened. Being realistic about your expectations plays a big part in helping to avoid this, but so does being sensible about the early events that you enter. Let's face it, if you've just bred your first batch of Orloffs or Cochins, and you decide to enter them in one of the more prestigious events in the poultry show calendar, then you're unlikely to get even a sniff of an award. However, sign them up for your local club event, or even at the chicken competition at your local agricultural show, and things might very well be different. Experience is everything when showing chickens, so start small and learn without too much pressure.

TIP 454: Take time to study the form

To get anywhere in a poultry show, the bird you enter will need to be a recognized pure breed, and from a strain that matches the official breed standard as closely as possible. The breed standards are overseen and approved by the American Poultry Association (APA) and the American Bantam Association (ABA), in conjunction with the individual breed clubs. Almost all properly organized poultry shows are covered by these fundamental rules. If you're not quite sure about whether your birds actually fit the bill, then get in touch with its breed club, or the APA, or do some research on the internet to clarify the situation. Also, if you want to enter an APA-affiliated event, then you'll have to join the club first. (For website details see page 288.)

TIP 455: *Pet chicken classes are a great first event*

The only real exception to the pure breed requirement at poultry shows can be found at some smaller shows, where the organizers recognize the wider popularity of the hobby, and allow the classification of "pet birds." You may also find junior classes that will accept entries with no restrictions. These will be judged much more informally and probably won't be a recognized part of the actual poultry event. Nevertheless, they can still be great fun to enter, and offer a good, basic grounding for younger enthusiasts into the fundamentals of the showing process.

TIP 456: *Whatever the show, effort's required*

Whether you're entering a pet competition at your local county fair, or have a bird in APA state, district, or annual meets, then it's important that you take the time to prepare it properly beforehand. This is a good habit to get into right from the start of your showing career. All competition entries will need to be clean, healthy, and free from parasitic infestation (a fault that'll see the bird disqualified from the show), otherwise the chances are that they won't be allowed into a show pen in the first place. Dirty birds simply aren't acceptable.

TIP 457: *Don't give yourself too much work*

While some of the most popular showing breeds include the Silkie, Rhode Island Red, Cochin, Orpington, Leghorn, Japanese, Australorp, Old English Game, and Marans, the one you choose for your exhibition debut can make a big difference to both your chances of success and the amount of work you'll have to do. Some breeds are much more demanding than others in terms of meeting the breed standard and show preparation. So, picking a relatively straightforward option, such as the Rhode Island Red or the Australorp, will make life much simpler. The relatively plain feathering and lack of additional complications like a head crest, intricate coloring, or feathered legs, mean that producing a presentable example is more of a realistic possibility for the novice breeder.

TIP 458: *Take time to get the paperwork right*

Any properly organized poultry show will produce an entry form, which has to be correctly completed by a certain date. These forms can appear a little daunting at first glance, as you'll need to provide full details of the bird being entered, together with the class in which it's going to compete. If you're unsure about doing this, then talk to the secretary or organizer of the event, who should be happy to help and advise. Alternatively, speak to another exhibitor for the dos and don'ts of poultry show entry and etiquette. This is another reason why it's sensible to begin your showing exploits at a local level, among enthusiasts you know.

TIP 459: *Know your breed categories*

Chickens in an APA show are classed a number of ways; the first category is size, which is broken into large fowl (standard size) and bantams (small size). Large fowl are further classified into categories named after the area of origin: American, Asiatic, English, Mediterranean, or Continental. Bantam classes include physical characteristics: Game, Single Comb Clean Legged, Rose Comb Clean Legged, all other Comb Clean Legged, and Feather Legged. Each class will be further broken into Breed and Variety, which includes coloring.

TIP 460: *Don't be late completing your entry*

Once you've completed the entry, get it posted off to the organization in plenty of time to meet the deadline. Ideally you should aim to get your details submitted three or four weeks before the closing date, so that your entry doesn't get caught up in the inevitable, last-minute rush, and run the risk of being left out altogether. A common mistake that many beginners make is forgetting to include the entry fee with the entry, so make sure you don't fall into that trap. The cost does vary from event to event, but it'll never be expensive; showing chickens remains an aspect of the hobby that's open to all thanks to its affordability.

GOOD PREPARATION

TIP 461: Good show preparation is a skill in itself

Selecting the right bird in the first place is a key factor, but knowing how to make it look its best is an art in itself. All too often novice exhibitors are pleased and satisfied with the way their birds look until they get to the show hall and see the appearance of the opposition. It can be disheartening at first, but the only way you'll improve the show presentation of your birds is with practice. If you're at all serious about exhibiting your birds, and putting them in front of a qualified poultry judge as well as the visiting public, then your show preparation must be thorough and effective.

TIP 462: It is important to choose the right bird

Selecting the birds that you think will be suitable for showing is one of the most important parts of the process, as well as one of the most difficult. The bird or birds that you're considering need to be as near to the breed standard as possible. They must present a good overall appearance, be able to walk properly, and show a full set of unbroken and correctly colored and marked feathers. Plumage-related problems are a very common cause of failure in the show pen. Even one broken flight or tail feather can be enough to prompt a judge to overlook that particular bird in favor of one with a complete and undamaged set.

TIP 463: *Watch your bird's weight*

Another important aspect to check before the show is the weight of the bird you're entering; it needs to fall within the limits set out in the official breed standard. While a good number of judges don't bother with actually weighing any more, some still do, so be warned. Also spend time checking over the fine details. Handle the bird carefully and search for any subtle defects that might catch the judge's eye: things like a missing toe nail, a twisted breast bone, knock knees, a crossed beak (see tip 328), or an incorrect undercolor. Then, once you're satisfied that the basics are right, make a final assessment of the bird's relevance to the breed standard, in terms of how well it matches with regard to leg, beak, and eye color and so on.

TIP 464: *Ensure your bantams are actually bantamweight*

One of the interesting aspects of poultry exhibition is how fashions change, and how this triggers different attitudes in the way standard-bred birds are judged. Now, you might well imagine that, with specific breed standards available in every case, assessing whether or not a bird conforms —especially with regard to a precise factor such as body weight—is relatively straightforward. However, increasingly, many judges prefer to assess weight visually. This, somewhat inevitably, has led to a gradual increase in size of many bantam types, sometimes almost to the point where it becomes difficult to be sure whether or not they are actually large fowl! For your own peace of mind, though, the best advice is to follow the standard.

TIP 465: *Exhibition birds need training for the big day*

Pen training your show bird really is the first step in the show preparation process. Essentially it involves acclimating it to the show pen environment. It's important that it becomes relaxed and calm when in this restricted space, so that it's not flapping around like a wild thing while being judged. It can take a few weeks to get a bird used to this level of confinement, especially if it's been leading a free-range lifestyle. The hen will also need to be handled on a regular basis so that it becomes comfortable with this; the judge will need to do this as he or she makes their assessment and, once again, birds that are hard to handle tend to get overlooked come judging time.

TIP 466: *Pen training can take quite a while*

You'll need to find yourself a pen that's a similar size to the sort commonly used at exhibitions. If you can't borrow or buy one of these wire units, then an alternative option is to build a simple, wooden-framed, hardware cloth version of your own. It's certainly worth taking a bit of time over this if you think you'll be showing a lot, as you'll be using it again and again. The time a bird spends inside the cage should be very short to begin with—just 10 or 15 minutes—but can gradually be increased every day. Continue like this until you reach the point at which the bird appears completely relaxed with the confinement. Make sure you provide the small cup waterers and feeders inside the cage, so that the bird gets used to these too. Unfortunately, some birds never become used to the restrictive space, so you simply have to give up and move on to another.

TIP 467: *Be ready for an early start*

Taking part in all but the most local of poultry shows is likely to involve an early start; the birds will have to be penned and ready for the judges reasonably early in the morning—usually by about 09.30 a.m. (details of this will be included in the event schedule). However, you'll need to be at the venue at least an hour earlier to get yourself sorted out and, more importantly, have the birds in place and settled. Attending a show can be a stressful time for chickens; they're out of their element and inevitably feel uneasy both on the journeys to and from the show and while at the event. It's important that you do all you can to minimize the effects of this upheaval by staying calm and quiet and working slowly and with care at all times.

TIP 468: *Keep show birds inside for a week before a show*

Birds that are badly presented, or have obvious signs of defects or health issues will, in most cases, be disqualified. The problem, of course, is that chickens just love dirt. The daily routine of those enjoying a free-range lifestyle will typically involve dust baths, scratching among bushes, and digging in the compost pile. Consequently, they quickly become grubby, especially if the weather is wet, which means that any birds you have earmarked for showing must be brought inside at least a week before the event to begin their visual transformations.

TIP 469: *Take great care when washing your bird*

The idea of washing a chicken may seem a little odd, but it really is essential for show birds. A full body wash is recommended to make sure the bird is really clean and lice free; insects and parasites are much easier to spot when the bird's plumage is wet. Many exhibitors wash their birds using medicated cat or dog shampoos, which are gentle enough not to cause problems but, nevertheless, still provide a thorough and effective cleaning action. The key to a good washing process is to remain gentle and calm with the bird at all times, to minimize stress; this whole process can be traumatic if done unsympathetically. The water used should only ever be just warm (never hot), and you must take great care not to splash water or cleaning product on to the head. Generally speaking, it's best to stand the bird in a shallow-filled bowl or sink and, if possible, to use a spray-type attachment to wet and then rinse the feathers.

TIP 470: *Use the bird's behavior as your guide*

If you don't have a spray head (or mixer tap), then use two bowls, with the second containing clean, warm water that's reserved for the rinsing stage. Never press hard or squeeze roughly when washing a chicken, and keep an eye on its behavior. If you notice it holding its beak open and/or gasping, it's stressed and you should stop for a few minutes to let it calm down. Once the washing stage is complete, wrap the bird gently in a clean, dry towel and leave this in place for a few minutes to soak up most of the surface water from the feathers. Then use a hair dryer set on its lowest heat setting (never on "hot") to finish the drying process. Alternatively, if the weather's warm and dry, you can leave the bird inside to dry naturally.

TIP 471: *Never allow a wet chicken to become chilled*

The big advantages of using an appropriately set hair dryer after washing are that it'll speed the feather-drying process and eliminate the risk of the bird becoming chilled. Wet chickens can quickly become cold if they're not dried effectively, and this is something to be avoided. For this reason, it's not advisable to put birds outside immediately after washing, however suitable you feel the weather may be. With the bird thoroughly dry, settled, and comfortable, check again for any signs of parasites, and then treat with an appropriate antimite spray or powder as an additional safety measure. If using a powder, make sure that it's worked well into the feathers, and penetrates right down to the skin.

TIP 472: *Allow time after washing for oil replacement*

Once your prized exhibit has been thoroughly washed, it'll require a few days to replace the lost oil from its feathers (removed during the washing stage). Obviously, you don't want the bird free ranging during this period and undoing all your careful cleaning efforts, so keep it confined to a small pen or show cage. Not only will this ensure that it stays completely clean but, being in isolation, there will be no risk of feather-pecking damage either. Then your thoughts must turn to the manicure stage of the preparation process. The bird's beak and nails will need to be checked, cleaned and, if necessary, trimmed. Take great care with any cutting; avoid over-cutting as this is likely to cause bleeding and discomfort. Afterwards, the beak and nails can be finished with a nail file to smooth away any jagged edges, which not only look unsightly, but can promote splitting too.

TIP 473: *Use an old toothbrush to clean leg scales*

All chickens have scaly legs, and these can get particularly engrained with dirt on birds that spend a lot of time outside. While you may well have removed any loose dirt during the washing stage, the chances are that the legs will still need a good scrub to look their best. An old toothbrush is just about the ideal tool for this job, and is best used with soap and water. Don't forget the feet and toes, both on the top and underside (a good judge will always check under here).

TIP 474: *The head is the natural focus of attention*

The last preparation job concerns the head. This is always the natural focus of attention for anyone viewing the bird, so it's very important that it looks its best. Pay particular attention to the beak, face, comb, ear lobes and wattles. All must be clean and bright to give the right impression. Gently clean these areas with a soft cloth and warm, soapy water, then rinse carefully with fresh, warm water once all traces of dirt have been removed. You can add a light dressing of baby oil to the comb and wattles to emphasize their color and create a pleasant sheen at this stage if you wish, or leave this final touch until the day of the show. Speaking of last-minute sprucing up, some old hands like to use a piece of silk for the final smoothing and buffing of the feathers, once at the show venue.

ON SHOW DAY

TIP 475: *Always transport your birds with care*

When taking your birds to and from a show it's important to consider how they travel. You'll need a suitable carrier that's both sturdy and offers plenty of ventilation for the occupant inside. Transport boxes have plenty of clean softwood shavings on the floor, to help the bird remain comfortable and clean during the journey. Box size is an important issue too. Never cramp the bird and always make sure that there's enough headroom for it to stand comfortably if it wants to. Boxes that are too small will be uncomfortable for the bird, and may also lead to damaged feathers. Remember to keep your car interior fresh and reasonably cool, and never be tempted to put water inside the carrying boxes while on the move.

TIP 476: *Make sure your entrant gets to the right pen*

Once you and your bird reach the show hall, the first job is to get it out of its traveling box and into the exhibition pen. For this you'll need to refer to your penning slip to find out which pen your bird has been allocated. Sometimes these will have been posted before the event, but often they'll be available for collection on the day. If yours hasn't already been sent to you, then ask any of the stewards for help (it's what they're there for) and they'll guide you to the table where the penning slips will be awaiting collection. It's obviously important to make sure that your bird finds its way into the correct pen. This procedure may vary, so check with the show organizer or coordinator beforehand.

TIP 477: *Delay providing food and water until after judging*

Once you have your bird safely in its correct pen on the morning of the show, don't be tempted to give it any food or water at this stage. It should not really be necessary assuming you provided both for the bird before you left home, and it'll greatly increase the risk of unnecessary mess at a critical stage. The judging should take no more than a couple of hours, after which you can make sure that food and fresh water are provided right away. Incidentally, it's considered bad form to interrupt the judge while he or she is going about their task. Any questions you may have should wait until after the decisions have been made. If the decision doesn't go your way, by all means ask about the reasoning but don't challenge the result. Accept the verdict with good grace, learn from it, and move on.

TIP 478: *Remember, the judge's decision is final*

The judge's decision at a poultry show is always final, and there's an unwritten rule that questioning the wisdom of his or her choice is bad exhibition etiquette. This, of course, doesn't stop some irate owners arguing with the officials, but all this achieves is unnecessary ill will and the generation of a troublesome reputation for the arguer. Similarly, it's very important that you don't approach or touch your bird while the judging is taking place. In many shows the hall is cleared of both exhibitors and the public during judging, although this isn't always the case. So keep your distance to avoid any untoward accusations that may be leveled after the event; judges are not supposed to know who has entered which bird. Once the judging has finished, and the prize cards are in place, you'll be free to tend to your bird as necessary.

TIP 479: *Don't be overeager to pack up and leave*

At the end of the day the event organizer will announce the closing of the show, and give permission for exhibitors to start "boxing" their birds. It's important, from a security point of view, not to remove your birds from their cages before this announcement is made. Bird crime from shows does, regrettably, seem to be on the increase, which is one of the primary reasons why boxing is only permitted at a fixed time. Once the birds have been packed, and everyone's satisfied that all are accounted for, then you'll be allowed to leave the venue and head for home.

TIP 480: *Postshow isolation is an important precaution*

When you get home from a poultry event, don't forget to keep your show birds separate from the rest of your flock. This "quarantine" period—lasting up to 14 days—is very important from a disease point of view. There's no telling what your bird may have encountered at the event, so it's best not to take chances. These days of isolation will give you time to monitor general health, behavior, and the condition of droppings, without putting the rest of your stock at risk. Then, if after two weeks all seems well, you're clear to return things to normal.

EXHIBITING EGGS

TIP 481: *Most people follow a typical poultry progression*

As chicken-keeping experience grows, so does the interest in raising birds to meet the established breed standard. Of course, the only way to gauge the progress you're making—not to mention the quality of your birds and their eggs—is to enter a exhibition and pit yours against others of the same breed. Taking this step, however, can be a daunting one for the novice showman. Presenting your bird's eggs for the inspection of a critical judge, in the relative formality of an exhibition hall, can be nerve racking enough to put some people off altogether. Persistence and a willingness to get involved are key.

TIP 482: *Exhibiting eggs is a great way to join the world of fancy poultry*

If you find the idea of cutting your exhibition teeth in the bird classes a little overwhelming—and plenty of people do—there's an alternative and less daunting route into exhibition that many novices find far less of an ordeal: showing eggs. This offers a great way to find your feet in the poultry exhibition arena. You may be able to find a spot in a show with your local club. Before you move ahead, check the guidelines that strive to ensure that eggs are shown at a set weight, size, quality, and standard. The standard itself determines an egg's shape, size, texture, color, freshness, bloom, and uniformity.

TIP 483: There's more to eggs than meets the eye

There are many aspects relating to a fresh chicken's egg. Show guidelines are designed to provide the framework from which judgments can be made, allowing eggs entered in various classes to be meaningfully assessed. Key areas of judgment may include freshness, shape, and size. The freshness is of the utmost importance (eggs are cracked open on to a white plate to judge this) and old, stale eggs will be disqualified. Overall shape is primary too: a good show egg should have a greater length than width, and a good dome at one end. The bottom of the egg mustn't be too pointed. The overall shape should present a smooth and even curve.

TIP 484: Size and weight matter when it comes to show eggs

The size, while certainly an important factor, may not be one of the main judging points for competition eggs, although they will need to be of an appropriate size for the breed that they're from. However, the overall weight may be a focus, and there may be very specific limits within which particular types of egg must fall. Both over- and underweight eggs may be disqualified or penalized in certain sections. All should be weighed by the judge as part of the assessment process. The normal weight for a pullet/hen egg is 1¾oz (49.6g) to 2oz (56.7g). However, this may be exceeded after several months of production and after the molt. Bantam eggs should not weigh more than 1½oz (42.5g).

  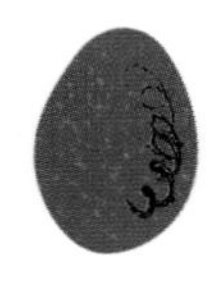 

TIP 485: *With eggs, don't take the rough with the smooth*

The shell of a good show egg should be smooth and free from any marks, bulges, or other obvious faults. The ends should be smooth with no roughness, porous areas, or pimples on any section of the shell. Shell color is an important consideration too. All manner of attractive colors are possible—white, brown, light brown (tinted), cream, blue, green, olive, mottled or speckled, and plum—depending on the breed of hen that's produced them. What's inside, of course, is just the same regardless of shell color. The yolk is the part of the egg that carries the germ cell from the mother bird, and it's surrounded by the clear albumen (white).

TIP 486: *Eggs entered in a poultry show must be fresh*

As an egg ages, it loses water and, therefore, weight, and the contents start to change. An egg judge will first test for freshness by gently shaking the egg, listening carefully to the sound it makes. Next, the egg is broken on to a white plate so that the contents can be carefully inspected. The size of the air sac will give a good indication of the age of the egg, as will the state of the contents. The clear albumen changes with age, appearing slightly cloudy when very fresh. However, with age its becomes increasingly thin and watery. Other signs of staleness include a yolk that looks flat and doesn't sit on top of the white.

TIP 487: *Air sac size is an indicator of freshness*

The air sac is always found at the blunt end of the egg, and can usually be observed using a candling lamp. It's formed when the inner and outer membranes separate soon after the egg is laid, and it acts as an air reservoir for the chick to use immediately before hatching, by which time it's expanded to account for about 40% of the egg's volume. In a fresh hen's egg, the air sac should only be about ¼in (5–6 mm) in depth or ⅝in (16–17mm) in diameter, increasing in size with age as the contents evaporate. When the diameter of the air sac is around ¾in (20mm), the egg can be regarded as being stale.

TIP 488: *White quality will give the freshness game away*

There are two distinct parts to an egg's albumen; the white immediately surrounding the yolk is thicker and should appear raised and firmer in a good, fresh egg. Then, surrounding this there should be a small amount of less viscous (thinner) white. When an older egg is broken on to a plate, the white will have degraded somewhat and, having become generally runnier, will spread further. You can test the freshness of an egg easily and simply at home. All you have to do is pop it into a bowl of water and watch what happens. If it's very fresh then it'll sink to the bottom and lie on its side. But, if it floats (due to the presence of a larger air sac), then this is an indication that it's no longer in its first flush of freshness. Despite its simplicity, though, this assessment method isn't used at a show.

TIP 489: *It takes expertise to be an egg judge*

There's a real art to being an egg judge. The primary objective is to achieve a level consistency in the decision making, which ensures that all egg entries are judged to a uniformly high standard. Those aiming to reach the standard must demonstrate good knowledge and understanding of everything from egg quality and appearance to contents quality, plus the finer points of egg decoration.

TIP 490: *Egg decoration is an option for the show novice*

Shows may offer the opportunity to exhibit painted, decorated, and displayed eggs. Some possible classes: painted eggs and decorated eggs. The first class includes blown or hard-boiled eggs that have been painted with oils, watercolors, ink, or any other media, without adding any other type of decoration such as beads. These can be as simple or elaborate as you wish. Check your show's guidelines, but the decoration may incorporate the use of beads, cardboard, paper, and the like—anything that can be used to enhance the overall effect.

HEN RESCUE

TIP 491: *Rescued hens can make great backyard pets*

A growing number of people are opting to adopt commercial hens from chicken farms, to save them from early slaughter. This kind, humanitarian act is typically driven by a compassionate wish to see birds that have been worked hard for 18 months under pretty intensive conditions enjoy a secure and relaxed retirement in a caring, domestic situation.

TIP 492: *Rescuing commercial hens isn't for everyone*

It's important to appreciate the responsibility attached to rehoming hens that have come from a commercial environment. While all chickens demand high standards of care and attention from their keepers, former "working" birds add an extra level of potential complication in that there's a delicate recuperation period involved with them too. Consequently, any responsible organization that's overseeing the rehoming of these birds will carry out some degree of vetting, in an effort to establish that their new home is going to be up to snuff. So applicants need to be prepared to meet the required criteria. After all, the last thing anyone wants is for these hens to be thrust into an unsuitable situation where they're neglected or possibly even ill treated.

TIP 493: It's easy to underestimate the trauma involved

Although the "rescuers" have the long-term good of the hens at heart, it's important to understand just how traumatic the whole rescue process is for the birds involved. Being removed from the only environment they've ever known is a terrifying experience for them. Their entire lives have been spent enclosed in secure, warm barns with carefully controlled, artificial lighting and nothing to do but eat, drink, and lay eggs. Swap this for the "alien" surroundings of a leafy backyard in suburbia, with grass under foot and limitless sky above, plus weather and strange noises all around, and you'll start to appreciate the enormity of the change. Chickens, being relatively delicate creatures when it comes to stress-inducing factors like changes in routine, can need coaxing through this transition.

TIP 494: There's no quick fix for a threadbare appearance

Contrary to what many people imagine, most hens living in the typical commercial, egg-laying environment are completely healthy; sick birds simply won't perform well enough. However, their pale, threadbare outward appearance can be a shock to the casual observer. Sparse feathering, washed-out comb and wattle coloring, and pale eyes tend to give a bad impression. What they also typically lack is muscle fitness, so they can be unsure as they move and prone to bruising in the stages after "rescue." Responsible rescuers should not rehome any birds with suspected health issues, but this doesn't mean that good health and trouble-free ownership thereafter is guaranteed.

TIP 495: *Take things very slowly for the first few days*

Once you get your rescued hens home, they should be placed carefully into the hen coop with as little fuss and noise as possible. Keep other pets and excitable children well away at this important stage. Don't encourage them out into the great outdoors; let them venture outside as and when they're ready to do so. It's important that they don't feel pressured to get out of their coop, or vulnerable once they are out. Secure them and their coop in a small run area while this acclimatization process is taking place, and keep them isolated from any other stock you may have.

TIP 496: *Be prepared to teach rescued hens the basics*

The first few weeks with commercial birds are likely to be a learning curve for you and them. You'll more than likely have to encourage them back into their coop at dusk, or if it rains heavily. Having been used to artificial light for 18 hours a day, it'll take them a little while to become familiar with the everyday routines of the real world. The hens may well prefer to sleep on the coop floor initially; don't be tempted to lift them on to the roosting perches at this stage. They may not be strong enough to withstand the jarring shock of jumping off the roost in the morning (bruising and broken bones are a risk here). So give them time to gain strength, and they'll start to roost when they're ready.

TIP 497: *Never be afraid to ask for help or advice*

One consequence of the upheaval associated with commercial hen rescue is that rehomed birds tend to be something of an unknown quantity. Some will recover strongly from the ordeal, blossoming into active and productive members of a backyard flock. Others, though, can find the stress of the whole ordeal too much. Some will die for no apparent reason and others may never lay another egg. Seek out the advice of others who've rehomed commercial layers for tips and pointers.

TIP 498: *Think carefully about hen coop location*

As far as housing commercial hens is concerned, all the same rules apply as for any other sort of domestic chickens. Apart from the careful management needed during their acclimatization period, the rest of their housing needs should hold no surprises. You'll have to provide secure, dry and draft-free accommodation for these birds, together with a predator-proof enclosure. One thing to bear in mind is that, having spent every previous living hour under cover in a threat-free environment, these birds are likely to be easily spooked compared to traditionally bred and reared chickens. For this reason, the location of their housing might benefit from being out of the way, to ensure a peacefully undisturbed lifestyle.

TIP 499: Think about what to feed rescued hens

You should seek a balanced and nutritious feed that will help promote feather regrowth and build strength and muscle tone at a sympathetically speedy rate. Ask your local feed source for advice on what to choose. Multiple feeding stations are a good idea during the early weeks too, just to make sure that all get their fair share without any bother.

TIP 500: Integration with existing birds needs a gentle touch

It's essential that rescued hens are given at least two weeks to recover both their strength and confidence in undisturbed surroundings. If you intend to mix them with an existing flock in time, then allowing the two groups sight of each other for a while before they finally meet can be advantageous. There's always going to be a degree of squabbling as a new pecking order is established; that's unavoidable. But careful observation and management should ensure that things don't get out of hand. If they don't get on, then bring out the commercial birds for their own safety. Plenty of keepers prefer to keep these birds permanently isolated anyway.

Popular chicken breeds

BREED	NO. OF EGGS	COMMENTS
Ancona	120	*Attractive, tough, disease-resistant, males can be feisty*
Andalusian	150	*Very attractive looks, reasonable layer but flighty*
Appenzeller	180	*Interesting looks, good layer, flighty, needs space*
Araucana	250	*Blue eggs, relaxed, nonsitter, good flier*
Australorp	200	*Good layer, family-friendly, ideal starter breed*
Barnevelder	180	*Great multipurpose, dark shells possible, increasingly rare*
Belgian bantam	50	*Tiny, active, decorative, can be aggressive, flier*
Booted bantam	80	*Manageable, pretty, great character, garden friendly*
Brahma	110	*Impressive looks, docile, easy to breed, getting rare*
Campine	180	*Decent layer, hardy, flighty, difficult to handle*
Cochin	110	*Cuddly giant, profuse feathering, easy to look after*
Dorking	120	*Docile, excellent table bird, ancient roots, five-toed*
Dutch bantam	165	*Surprisingly hardy, friendly, good flier, feisty males*
Faverolles	180	*Docile, attractive, decent table bird, needs space*
Fayoumi	220	*Great layer, hardy, attractive feathering, flighty, noisy*
Frizzle	160	*Odd feathering not to everyone's liking, hardy, rare*
Hamburgh	210	*Stunning looks, great layer, challenging to show*
Houdan	150	*Unusual crested looks, docile, utility roots*
Ixworth	210	*Good layer and table bird, fast grower, rare*
Japanese	50	*Attractive, many colors, friendly, tricky to breed*
Jersey Giant	230	*Great layer, docile, hardy, best given space*
Langshan	200	*Good layer, calm, friendly, hardy, great family bird*
Leghorn	230	*Excellent layer, easy to keep, long-lived, flighty*

BREED	NO. OF EGGS	COMMENTS
Malay	35	*Tall, impressive, placid, need plenty of space*
Marans	220	*Excellent layer, dark-shelled eggs, hardy, straightforward*
Minorca	190	*Good layer, friendly, easy to keep, disease-resistant*
Modern Game	100	*Stunning appearance, easy to keep, risk of theft*
Norfolk Gray	230	*Hardy, pleasant character, good table bird, very rare*
Orloff	50	*Hardy, quirky looks, increasingly difficult to find*
Orpington	60	*Very attractive, friendly, easy to keep, disease-resistant*
Plymouth Rock	180	*Non-flier, friendly, decent layer, easy to breed*
Polish	120	*Great looks, good color choice, high maintenance*
Redcap	175	*Hardy, attractive, decent layer, flighty, challenging*
Rhode Island Red	180	*Good for meat and eggs, docile, beginner's favorite*
Rosecomb	150	*Great looks, affordable, not hardy, short lived*
Scots Gray	150	*Great all-rounder, easy to keep, only one color*
Sebright	45	*Stunning looks, friendly, hardy, flier, difficult to breed*
Silkie	90	*Unique appearance, great broody, Marek's disease risk*
Spanish	170	*Large eggs, good looks, hardy, flier, very rare*
Sultan	80	*Amazing looks, friendly, high maintenance, hard to breed*
Sumatra	120	*Stately looks, docile, good broody, somewhat flighty*
Sussex	220	*Great multipurpose, easy to keep, roosters noisy*
Transylvanian	150	*Bizarre looks, easy to keep, hardy, good flier*
Welsummer	140	*Pretty bird, exotic dark eggs, hardy, reasonable layer*
Wyandotte	175	*Easy ownership, useful table bird, large fowl scarce*
Yokohama	150	*Spectacular long feathers, good mothers, aggressive males*

INDEX

USEFUL WEBSITES

AMERICAN POULTRY ASSOCIATION
www.amerpoultryassn.com

CORNELL UNIVERSITY ATLAS OF AVIAN DISEASES
http://partnersah.vet.cornell.edu/avian-atlas

GREAT LAKES POULTRY ASSOCIATION, ONTARIO, CANADA
www.greatlakespoultryclub.ca/

HEALTH CANADA
www.hc-sc.gc.ca/cps-spc/pest/index-eng.php

HENDERSON'S BREED CHART
www.ithaca.edu/staff/jhenderson/chooks/chooks.html

SOCIETY FOR THE PRESERVATION OF POULTRY ANTIQUITIES (SPPA)
http://sppa.webs.com

UNIVERSITY OF KENTUCKY, LIST OF BREED ASSOCIATIONS AND CLUBS
www2.ca.uky.edu/smallflocks/links/breed_associations.html#US